I0479074

TRANSACTIONS

OF THE

AMERICAN PHILOSOPHICAL SOCIETY,

HELD AT PHILADELPHIA,

FOR PROMOTING USEFUL KNOWLEDGE.

VOL. XVIII.—NEW SERIES.

PUBLISHED BY THE SOCIETY.

Philadelphia:
MacCalla & Company Inc., Printers,
1896.

CONTENTS OF VOL. XVIII.

CORRECTIONS.

Page 9, l. 29 : For *Allen* read *Allan*.

Page 14, l. 31 : For *lissukâ* read *lissuhâ*.

Page 15, l. 24 : . . . . . is to be removed.

Page 29, l. 26 : For I read II.

Page 37, l. 5 : For *Barnaburiash* read *Burnaburiash*.

Page 43, l. 26 : For *Ménaut* read *Ménant*.

TRANSACTIONS

OF THE

AMERICAN PHILOSOPHICAL SOCIETY.

ARTICLE I.

OLD BABYLONIAN INSCRIPTIONS
CHIEFLY FROM NIPPUR.

BY H. V. HILPRECHT, Ph.D.,

Professor of Assyrian and Curator of the Babylonian Museum in the University of Pennsylvania.

Read before the American Philosophical Society, November 4, 1892.

PREFACE.

THE old Babylonian Cuneiform Texts, which are published in the following pages, are a part of the harvest gathered by the Expedition sent out in the summer of 1888, under the auspices of the University of Pennsylvania, for the exploration of Babylonia. The Rev. Dr. John P. Peters, Professor of Hebrew in the University of Pennsylvania, was the Director of the Expedition, while the subscriber, as the Assyriologist of the University, accompanied it during the first year of its labors. As the history of the Expedition is to be published by its Director at an early date, I here abstain from giving any account of its origin, members, undertakings and results. In the meantime for the student I have appended to the Introduction a Bibliography of those contributions of its members to various periodicals which relate to its work.

Towards the close of the year 1891 there arrived at the Museum of the University some eight thousand clay tablets, together with several hundred fragments of vases and other inscribed objects in stone, which had been disinterred in Nippur or Nuffar.* I was able at once to proceed with the work of cleaning and examining

---

* This is the present designation of the extensive ruins by the Affek tribes, in whose territory they are situated. Although I repeatedly had the Arabs of the neighborhood pronounce for me the name they give to the ancient Nippur, I never heard from their lips the pronunciation Niffer, to which Layard and Loftus have given currency among Assyriologists.

them. Three months later I had obtained a general idea of their contents and their age, and had catalogued about a third of them. On the basis of a report submitted to the Publication Committee of the Expedition, of which Mr. Clarence H. Clark is Chairman, a plan was carefully devised for making these cuneiform inscriptions accessible to a wider circle of students, with as much speed and method as possible. With this view the Assyriologists of America and Canada were invited to lend their aid to the preparation of an extensive work on the Expedition and its results. A number of them have given assurance of their readiness to do so.

In April, 1892, the undersigned was entrusted by the Committee with the editing of the series containing the Cuneiform Texts, and, at the same time, was requested to undertake at once the preparation of the first volume of these texts. It is estimated that the series will extend to eight or possibly ten volumes. Their general plan and character are well explained in a report submitted to the American Philosophical Society by a special committee, of which Mr. Talcott Williams was the Chairman, at the stated meeting of May 20, 1892.

I take this opportunity to acknowledge the liberality of the venerable American Philosophical Society of Philadelphia, as shown in the promptness with which it has undertaken the publication of the present volume, by giving it a place in its learned and valuable Transactions. I hope that in the future the Society will continue to evince its interest in making such labors accessible to the republic of letters, by extending its sympathy and support to the undertaking whose plan has been described.

A word more must be said as to the manner in which it is intended to prepare the Cuneiform Texts for the use of the Assyriologist. For the sake of securing uniformity throughout the series, and of avoiding what would make it excessively costly, it was necessary to reproduce the inscriptions by photograph from copies made by hand, rather than from the objects themselves. Besides, the editor some time ago reached the conclusion that the method of direct photography is not at all satisfactory in the case of many inscriptions. The best which has been done by that expensive process is beyond question the work edited by Ernest de Sarzec and Léon Heuzey under the auspices of the government of France: *Découvertes en Chaldée*. It possesses unique merits. But in spite of all the care that has been taken to secure an exact reproduction of the monuments, any Assyriologist who has worked through such texts as are found on Plates 33, 35 and 41, No. 1, will agree with me that the decipherment, especially of the margins, makes a very severe demand upon the eyesight—a circumstance which makes the prompt and comprehensive use of the contents of this beautiful work sometimes difficult. After mature consideration, therefore, the Committee found it most suitable to reproduce the Cuneiform Texts from copies made by the hand, and to employ photographs from the objects themselves only occasionally, to enable the Assyriologist to verify the copies and to perceive the archæological character of the inscribed objects.

The first volume, whose first part I publish herewith, contains only inscriptions in old Babylonian which have been found on vases, door sockets, stone tablets, votive axes, bricks, stamps, clay cylinders, and similar objects of a monumental character. As the most of them belong to that period of Babylonian history of which our knowledge is very defective, the most painstaking care has been applied to autographically reproducing the originals with the utmost faithfulness. The editor has kept in view, not only the making fresh and important materials accessible to students of Assyriology, but also the doing his part in placing Babylonian paleography on a better foundation. For this end every text has been reproduced in its actual size and form—that is, so as to show all the peculiarities of the scribes, not only as to the dimensions, shape and position of every character and group of such, but also their distance from one another, as was so admirably done by Sir Henry Rawlinson and Edwin Norris in the first volume of *The Cuneiform Inscriptions of Western Asia*. The investigations and collections I have made since the year 1883, and my lectures regularly held since 1886 on "The Development of Cuneiform Writing in Babylonia and Assyria," have led me to conclude that the size and relative position of individual cuneiform characters, and certain combinations in which they frequently occur, have been a factor of importance in the development of the stereotyped forms of later date. The detailed proof of this I must reserve for the present until more urgent matters have been disposed of. At any rate, careful editions of texts, and a faithful reproduction of the peculiarities of the individual Babylonian scribe, have become a pressing necessity for the progress of Assyriology, if we are to attain in this field anything like the results which Euting has achieved in other departments of Semitic paleography, and which are so necessary in determining the age of fragmentary and undated inscriptions. In spite of the scantiness of representative old Babylonian texts of which the Assyriologists could make use, it would not have been possible for them to have differed by 500, 1000 or even 2000 years as to the date of inscriptions, if such texts had always been reproduced carefully for their use.

It is to be expected that the excavations still proceeding at Nuffar will supply the completion of texts here given in fragmentary shape, and that several finds will make their way into various European and American museums by reason of the thievishness of the Arabs employed in them, who also may carry on excavations on their own account.* For this reason I have shown as exactly as possible the fracture

* Cf. my note in *Zeitschrift für Assyriologie* IV, p. 282 *seq.* Sayce, *Records of the Past*[2], Vol. III, pp. x, note 3, XV.

of such fragments. It was thus that I myself, after the printing had begun, was enabled to recognize the connection of Pl. 21, No. 41 and No. 46, and between Pl. 22, No. 50, and Pl. 26, No. 74.

Where I have shaded the inscription in my copy, it is not meant to indicate that the reading is to me uncertain, but that it can be recognized only in a special light and by a practiced eye, looking at it from an especial angle. How necessary it was to make an autograph copy of such inscriptions may be seen by comparing Pl. 23, Nos. 56, 57, and the direct photographic reproduction on Pl. X. A restoration of broken characters and lines I have avoided on principle, even when there was no doubt in my own mind as to what was missing. My translations will show in due time what my understanding of such passages is. For obvious reasons, I have given the characters in some inscriptions only in outline. Of the plates which reproduce the inscription on the Abu Habba slab I have avoided altogether making an autograph copy, since I thought this needless. This stone was found in Abu Habba during the excavation undertaken at the private expense of the Sultan in 1889, and is now in the Imperial Museum at Constantinople. Through the courtesy of His Excellency Hamdy-Bey, a cast of it was furnished to our Expedition. Unfortunately this was broken in pieces in transportation, but it was restored by one of my students. It is this cast that has been directly photographed for the present publication. Some portions of its margin have an indistinctness, which is faithfully shown by the photographic reproduction.

To convey to scholars a clearer picture of the ruins of Nippur, and to show the sites at which the several inscriptions were found, a plan of the excavations of the first year is given. In the Table of Contents the texts are described with reference to this Plan, which has been prepared in accordance with the bas-relief of the ruins made by Mr. Charles Muret in Paris under the supervision of Mr. Perez Hastings Field, the architect of the Expedition.

In determining the mineralogic character of the several stones, I have had the assistance of my colleagues, Drs. G. A. Koenig and E. Smith, of the University of Pennsylvania, to whom I extend my thanks. As I was able to accompany the Expedition only during the first year, I am greatly indebted to my esteemed colleague, Dr. Peters, for much valuable information as to the sites in which objects were found, and for sketches and copies of a series of objects and inscriptions which he made during its second year. As the antiquities disinterred arrived in this country at long intervals, I found myself obliged to proceed with the help of casts, squeezes, electrotypes and Prof. Peters' notebooks, in order not to delay needlessly the publication of the Texts. This circumstance, however, prevented my determining at the outset the material of the whole volume. At the opening of each new box I found myself compelled to withdraw some pages and substitute others, until the commencement of the printing, in October of last year, made further alterations and a more systematic arrangement impossible. The second part of this volume, which will appear in about half a year, will furnish further inscriptions of kings who are already represented in the first. Nor will it be possible entirely to avoid this defect of arrangement in other volumes, so long as the excavations at Nippur continue to bring to light new inscriptions of the same rulers. If, however, we were to delay the publication of the inscriptions until the complete results of the systematic explorations of the ruin-heaps at Nippur were at hand, it would have been necessary, according to my careful calculation, to wait some twenty years, supposing that the excavations were pushed forward with a force of some hundred Arab workmen.

On account of its importance and its close connection with the class of Cassite votive inscriptions here published, I have included the cuneiform text on the lapis lazuli disc of King Kadashman-Turgu, which probably came from Nippur,[*] and is now in the Museum of Harvard University,[†] Cambridge, Mass. Prof. D. G. Lyon kindly gave me leave to publish this, and placed at my disposal a cast of the disc, for which he has my warmest thanks.

The transcription of the names of kings in the Table of Contents is the usual one. A new transliteration has been substituted only where there are sufficient grounds for departing from that formerly used. The texts in the main have been arranged chronologically, in the order of the Babylonian dynasties; yet where the better utilization of space seemed to justify this, and also, as already said, because it was impossible to obtain at the outset all the material of the present volume, I have departed from that order in a few instances. Nor have I attempted to distinguish between the inscriptions of Kurigalzu I and II, simply because, with the material now at our disposal, it is not possible to do so with any certainty.

Three other volumes of cuneiform texts are in preparation. The transcription and translation of the inscriptions here given are as good as completed, and will appear at an early date. From this translation I have excluded the Abu Habba slab and the two Yokha tablets (Plates VI–VIII). These latter are to be treated in connection with other tablets of similar character and contents. A translation of the former I

* Cf. Hilprecht, "Die Votiv-Inschrift eines nicht erkannten Kassitenkönigs," *Z. A.* VII, p. 318.

† Cf. Lyon, "On a Lapis Lazuli Disc" in the *Proceedings of the American Oriental Society*, May, 1889, pp. cxxxiv–vii.

propose to publish separately in the course of next summer, in coöperation with my esteemed colleague, Dr. P. Jensen, Professor in the University of Marburg.

In conclusion, it is but just that I should express here publicly my profound gratitude to Dr. William Pepper, Provost of the University of Pennsylvania, Messrs. Clarence H. Clark, E. W. Clark, W. W. Frazier, Charles C. Harrison, Prof. Dr. Horace Jayne, Prof. Allen Marquand, Jos. D. Potts, Rev. Dr. H. Clay Trumbull, Talcott Williams, Richard Wood, Stuart Wood, and to all the other gentlemen whose lively interest in the history and civilization of ancient Babylonia, and whose liberal and constant support, have made possible the thorough researches at one of the most ancient ruins of the world. *   That the publication of this first part of the results obtained by the American Expedition does not take place until nearly four years after it was begun, is due to the extraordinary difficulties it encountered, on both sea and land, through shipwreck near Samos, through the hostility of Arab tribes, through the burning and plundering of our camp, through the outbreak of malignant cholera in Babylonia, through the delay of the antiquities on their way to America, and through the severe illness from which nearly all the members suffered.   Often it seemed as though the grewsome curse of King Sargon I, one of the oldest monuments of Semitic speech published in the following pages, had rested on the American Expedition, as that of the Phœnician king Eshmunazar rested on Napoleon : " Whosoever removes this inscribed stone, his foundation may Bêl and Shamash and Ninna tear up, and exterminate his seed!" We trust, however, that the rage of Enlil, lord of the demons, who set loose against the Expedition all the Igigi and Anunnaki, will abate with the publication of these cuneiform inscriptions, almost every one of which proclaims the glory of the great Bêl, " lord of the lands," and that the curse of nearly six thousand years ago will be transformed into the kindly blessing which King Nazi-Maruttash utters in his poetic prayer:

ikribishu ana shemê	to hear his prayer,
teslissu magâri	to grant his supplication,
unnênishu lekê	to accept his sigh,
napishtashu naṣâri	to preserve his life,
ûmêshu urruke	to lengthen his days.

(Pl. 27, No. 78.)

H. V. HILPRECHT.

PHILADELPHIA, January 1, 1893.

* Cf. Pinches, *Records of the Past*², Vol. VI, p. 109, l. 6. (The Non-Semitic Version of the Creation Story).

INTRODUCTION.

THE cuneiform tablets and stone inscriptions, excavated by the Expedition in Nippur, embrace a period of about 3350 years—c. 3800 to c. 450[1] B. C. About one hundred and twenty kings of Babylon, Ur and other cities are known to belong to this period of Babylonian history. Forty-five of these, according to our present knowledge, have left personal inscriptions or documents dated according to their reigns in Nippur. Several of these rulers, whose names were only partly preserved or otherwise obscure, or whose chronology and duration of reign were doubtful, have been placed in new light by the American excavations, while others can now for the first time be studied from their own inscriptions. Among other points the following have been established: The correct reading of *Ur-Ninib* of Isin, instead of *Gamil-Ninib*[2] as heretofore; the proof of the existence of King *Ibil-Sin*, or better, *Îni-Sin* of Ur,[3] already discovered by George Smith,[4] but not generally accepted by Assyriologists; the proper pronunciation of the name *Nazi-Maruttash;*[5] the correct transcription of the group *Ka-dash-man*, instead of the hitherto *Ka-ara*, in a series of Cassite proper names;[6] the completion of the name of the twenty-seventh king in the Babylonian list b[7] to *Shagashalti-Shuriash*[8] (Shamash is deliverance), instead of the usual *Shagashalti-Buriash*[9] (Rammân is deliverance); the completion of the Cassite king [. . . . . . *i*]*a-shu* in *S.* 2106, Obv. l. 9,[10] to *Bibeiashu*, and the identity of the latter with *Bibe*,[8] the son of Shagashalti-Shuriash; the first inscription of the

---

[1] Contract dated in the reign of King Artaxerxes I. A number of coins, about one hundred terra cotta bowls bearing Hebrew, Syriac and Arabic inscriptions, and many other objects, which belong to the Nippur of the Christian era, are here excluded.

[2] Hilprecht, "Die Votiv-Inschrift eines nicht erkannten Kassitenkönigs" in *Z. A.* VII, p. 315, note 1.

[3] Hilprecht, "König Ini-Sin von Ur" in *Z. A.* VII, pp. 343–346.

[4] *Trans. Soc. Bibl. Arch.* I, p. 41.

[5] Hilprecht, *l. c.*, pp. 310, 311.

[6] Hilprecht, *l. c.*, pp. 309, 314, 315.

[7] Winckler, *Untersuchungen zur Altorientalischen Geschichte*, p. 146, col. ii, 6.

[8] Hilprecht, "Die Ergänzung der Namen zweier Kassitenkönige," *Z. A.*, in print.

[9] Cf. Winckler in *Z. A.* II, p. 310, and *Unters.*, p. 30.

[10] Winckler, *Unters.*, p. 152.

kings *Rammán-shum-uṣur*[1] and his son *Mili-Shikhu;*[2] and the determination of the approximate duration of the reigns of the Cassite kings *Kurigalzu, Nazi-Maruttash,* etc., their succession and kinship with each other. In addition, the following new kings have been added by the Expedition to those already known: 1. *Âlusharshid;* 2. *Bur-Sin I;* 3. *Gande;*[3] 4. *Kadashman-Turgu (Kadashman-Bêl);* 5. *Kudur-Turgu (Bêl);* 6. *Bêl-nâdin-aplu.*

Intending to give in the near future the transcription and translation of the inscriptions here published, I confine myself at present to the following points:

THE OLDEST SEMITIC KINGS OF BABYLONIA.

Of the cuneiform inscriptions of the oldest Semitic kings of Babylonia very few have been discovered. Winckler recently published them together in his *Altbabylonische Keilschrifttexte,* p. 22.[4] Undoubtedly to this ancient period belongs also the inscription[5] of the king of the country of Guti, *i. e.,* " of the country and people to the east of the lower Zâb, in the upper section of the region through which the Adhem and the Dijâlâ rivers flow."[6] Various reasons[7] compel me to differ from Winckler's determination as to the date of this inscription by about 2000 years, *i. e.,* to transfer it from the time of Agum (Winckler, *Geschichte,* p. 82), about 1600 B. C., back to the time of Sargon, about 3800 B. C.[8] Because of the very archaic form of the cunei-

[1] Hitherto represented only by a boundary stone dated in the time of the kings Rammân-shum-iddina, Rammân-shum-uṣur and Mili-Shikhu. Cf. Belser in *Beiträge zur Assyriologie* II, pp. 187–203 (quoted hereafter as *B. A.*) and Peiser in Schrader's *Keilinschriftliche Bibliothek* III, Part 1, pp. 154–163 (quoted hereafter as *K. B.*)

[2] For the reasons for identifying the king of the inscription Pl. 29, No. 82, with Mili-Shikhu, see below, p. 36.

[3] Unless identical with Gandash, the first king of the Cassite dynasty. Cf. pp. 28–30.

[4] Cf. Winckler, in Schrader's *K. B.* III, Part I, pp. 98–107.

[5] Published by Winckler, *Z. A.* IV, p. 406.

[6] Delitzsch, *Wo lag das Paradies?* pp. 233–237. Cf. Delattre, *l'Asie occidentale dans les inscriptions Assyriennes.*

[7] The predominant use of the archaic line-shaped characters, their marked agreement with a whole series of characters on Plates 1 to 5, the Semitic speech, and its whole phraseology, together with the peculiarities to be seen in the sibilants, which are the same in the texts of Sargon I from Nippur, the fact that Abu Habba, where other texts of the same high antiquity have been disinterred, is the place of its discovery, the use of a "perforated stone" as votive object for the inscription, itself a characteristic of ancient times, the mineralogic character of the stone, and last of all—just what Winckler (*Z. A.* IV, p. 406) is disposed to regard as proof of a later origin—the notably sharp and skillful carving of the inscription. This last proof is especially convincing, for it is a characteristic trait of the oldest Semitic cuneiform inscriptions carved in stone, that they are engraved with a beauty and a sharpness which are absent from those of later date (cf. also Hommel, *Geschichte,* p. 301).

[8] It will not be objected that the cuneiform characters, indeed, seem to indicate a great antiquity, but that they may very well be an imitation of the work of an earlier period by a later king. This has become a very favorite mode of reasoning when the date of an undated inscription is to be determined from its writing (*e. g.,* Amiaud et Méchineau, *Tableau Comparé,* p. xiii *seqq.,* Pinches, *Hebraica* VI, p. 57), and serves to produce a very chaos of uncertainty in the province of Babylonian paleography. I think it opportune to state here that I am not acquainted with

form characters and of certain mutilated passages, this inscription of the king of Guti presents great difficulties, so that, to my knowledge, it has never been translated, and Winckler has come to the conclusion that it was composed "apparently in part in the native tongue" of the king of Guti. Winckler would not be entirely incorrect if he understood by this "native tongue"[1] the Semitic-Babylonian of the inscriptions of Sargon I, for the text is written in pure Semitic-Babylonian, and reads as follows: 1. *La-si*(?)-*ra*(?)-*ab*(?) 2. *da-num*[2] 3. *shar* 4. *Gu-ti-im* 5–10. vacant 11. *ip-ush*(?) -*ma* 12. *iddin* 13. *sha duppa* 14. *shú-a*[3] 15. *u-sa-za-ku-ni* 16. *zikir shum-su* 17. *i-sa-da-ru* 18. *ilu Gu-ti-im* 19. *ilu Ninna* 20. *ù* 21. *ilu Sin* 22. *ishid-su* 23. *li-su-ḫa* 24. *ù* 25. *zêra-su* 26. *li-il-gu-da* 27. *ù* 28. *ḫarrân alkat*(*-kat*)*-su* 29. *a i-si-ir*, "Lasi-rab(?), the mighty king of Guti, ... has made and presented (it). Whosoever removes this inscribed stone and writes (the mention of) his name thereupon, his

a solitary instance in which such an imitation of the older cuneiform characters by a later Babylonian ruler has been shown with certainty. What is commonly regarded as such may be traced to a lack of carefulness in examining the single characters of the inscriptions in question. Gande's endeavor to imitate the characters of earlier Babylonian kings is to be judged entirely differently (see below). In Babylonia at all times two systems of writing—a hieratic and a demotic—existed side by side. The latter is the system used in the affairs of everyday life, and was subject to a continuous process of change and development, which resulted at last in the stereotyped cuneiform characters of the Neo-Babylonian and Persian contract tablets. What I have called the hieratic system of cuneiform writing was identical with the demotic in the earliest times; but later was confined to religious literature (including seal-cylinders) and formularies originally bearing a religious character (boundary stones, etc.). Although, in the nature of things, it was less subject to change than the other, yet it developed distinctly different forms of most characters in the different periods of its history. In more or less dependence upon the material inscribed, the local tradition and the peculiarities of the individual scribe, the hieratic writing also passed through a course of development, more limited in extent, but peculiar to itself. When due attention is given to these facts in every case, there will be an end to the weltering confusion of early and late texts, and of the critical helplessness which results from this, in the field of Babylonian paleography.

[1] It is true, indeed, that the question as to whether the earliest inhabitants of Guti spoke a Semitic language (cf. Hommel, *Geschichte*, pp. 279, 306, note 2) cannot be regarded as definitely answered, if we maintain that the "perforated stone" was a gift of the king of Guti to the temple in Sippara (cf. "The King of Châna," *Trans. Soc. Bibl. Arch.* VIII, p. 352). In this case the inscription might very well have been composed in the Semitic dialect used in Sippara. I hold, however, that the object was not a gift of the king of Guti to the temple of Sippara (observe the absence of god Shamash and the first position given to god Guti), but that it had been carried off as booty from the land of Guti by one of the earliest Babylonian kings, in the same way as the vase of Narâm-Sin (*namrak Magan*) and most of the vases of Alusharshid (cf. Pl. 4, l. 11, 12: *namrak Elamti*) were carried to Babylonia. From this it certainly would result that, just like the inhabitants of Lulubi (cf. Scheil, *Recueil de Travaux* XIV, livr. 1 et 2, p. 104), so also those of Guti spoke Semitic and worshiped the Babylonian gods Ninna and Sin, along with their principal national god Guti. This last deity seems to have given his name to their country, as did the god Ashur to the city and land of Ashur (cf. also Ni(a?)nna and Nineveh, etc.), and the god Shûshinak to the city of Shûshinak or Susa (cf. Hagen in *B. A.* II, p. 233).

[2] Cf. Jensen, in Schrader's *K. B.* III, Part I, p. 116, note 5.

[3] Winckler offers *za*. Apparently this reading results from an oversight either on the part of Winckler or of the ancient scribe; for cf. Pl. 1, 13; Pl. 2 (and I), 14.

foundation may Guti, Ninna and Sin tear up and exterminate his seed, and may whatsoever he undertakes not prosper!"[1]

To the time of Sargon and Narâm-Sin[2] belongs also the first of the two inscriptions of Ser-i-Pul (*Stèles de Zohâb*), published by Messrs. J. de Morgan and V. Scheil in *Recueil de Travaux relatifs à la Philologie et à l'Archéologie égyptiennes et assyriennes* XIV, *Liv.* 1, 2, 1892, pp. 100–106. Both of these badly mutilated inscriptions are written in a Semitic[3] dialect, and the phraseology is very similar to that of the king of Guti. Scheil offers a transliteration and translation of the preserved portions. In regard to the first inscription I remark, however, that col. I, 11: *ù* DUB BA AM, can hardly be read (with Scheil) *u dubbam*.[4] The preceding phrase, *ṣalmêtum annitum*, "these images," and the parallel passage of the Guti text and Pl. 1 and 2 of the present volume—*duppa shu'a*—require a demonstrative pronoun in connection with *duppa*. I therefore regard *BA* as the ideogr. for *shu'atu*,[5] and read *duppa shu'atam(-am)*, "this inscribed stone." The second character in col. II, 10, which Scheil does not recognize (*l. c.*, p. 105) is *il*,[6] and the line

---

[1] In the interpretation I remark the following: L. 2. *da-num* is not to be regarded independently as an appositive representing the usual *sharru da-num* (*Stèle de Zohâb* I, col. 1, 2), but must be joined with *shar Gutim*, as "the mighty king of Guti." The position of the adjective before the substantive is not so much due to the emphasis of the adjective (Del. *Gram.*, § 121) as to the endeavor to avoid separating the adjective from the noun to which it belongs. L. 14. *Shu'a* (or *shuwa*) is the older form from which *shu'atu*, resp. *shu'âtu*, has been derived. Cf. Arabic *huwa*, Del. *Gram.*, § 57, and Jäger, in *B. A.* I, p. 481 *seq.* L. 15. 17. *usazakuni*, *isaṭaru* are not present tenses of the stems III₂ and I₂ respectively (= *utsazakuni*, *itsaṭaru*), but, in consideration of l. 29, are to be regarded as III₁ and I₁ = *ushazakûni* (*Stèle de Zohâb* I, 12) = *ushazzakûni* = *ushanzaku* + *ni* (Del. *Gram.*, § 79 β) and *ishaṭaru*. *Sh* between two vowels, or with an *m* following, was apparently pronounced as *s* (cf. also *Pl.* 1 and 2). The root of *usazaku* is נוך or נוק, II R. 30, 42, *e, f* (Jensen, *Kosmologie*, p. 339), not מצך (Scheil, *l. c.*, p. 108). It means "to be in motion, to move" (intr.). Cf. *naziktu*, II R. 23, 65, *e, f*, synon. of *daltu*, "door" = "that which moves (on a hinge);" *izzuk mulmullu* (Creation Tablet IV, 101), "the spear quivered." III₁ = "to move (trans.), to remove." This meaning is supported by parallel passages, as V R. 33, col. VIII, 42: *mannu sha itâbalu* (Jensen, in Schrader's *K. B.*, III, Part I, p. 152, note 3) *shumishu kîma shumi'u ishaṭaru*, "Whoever carries off (the tablet) and writes his name as my name." L. 16. The sign *gish*—dialect. for MU—signifies apparently *zikru* (Sargon Cyl., l. 50). Cf. Jensen, *Z. A.* I, p. 184. L. 23. *lisuḫà* = *lissuḳâ*, נסך. Cf. Pl. 2, 20 (Pl. 1, 21: *lissuḫû*). For the *à* of the 3d pers. masc. plur., cf. Del. *Gram.*, § 90, c. L. 26. *li-il*(sic! = Brünnow, *l. c.*, 4847)-*gu-da* = *lilḳuṭâ*, cf. Pl. 2, 23. Pl. 1, 24 reads in its place *li-il-gu-tu* = *lilḳuṭû*, לקף. Cf. the corresponding Sumerian phrase at the close of the inscription of *Kadashman-Turgu*, Pl. 24, No. 63. L. 28 is uncertain. The second character I regard as DI = *alâku*, and the third character, *kat* (Brünnow, *List*, 2701), a phonetic compliment. According to the scribe's method of writing, we should expect but one word on this line. L. 29. *a isir* = *â ishir*, Præt. I₁ of ישר. Cf. III R. 61, No. 2, 14: *alkat mâti là ishshir*, "the business (*Handel und Wandel*, Del.) of the land may not prosper."

[2] Thus, correctly, Scheil, *l. c.*, p. 105. The second is considerably younger.

[3] Also the features of the king *Anu-banini* of Lulubi, carved together with the inscription in the rock, are manifestly Semitic.

[4] Scheil translates "*cette tablette*," but adds "cette" only from the general context.

[5] Perhaps it is to be read directly *shu*, and the two characters must be transcribed as *shu-am*. Cf. also Amiaud, in *Z. A.* II, p. 292.

[6] No. 73 in Amiaud et Méchineau, *Tableau comparé*, must be corrected accordingly.

reads *li-il-ku-du* = *lilḳutû*. The second inscription (*stèle de Cheikh-Khân*) is, in my estimation, misunderstood by Scheil. There is no question of "restoration," [1] but of the first erection of the image.

To this, the already known material touching the oldest Semitic period, has come now to be added Pl. 1–7. The above remarks upon the texts of the kings of Guti and Lullubi open the way for a better understanding of these new texts. The following notes supply all that still needs to be added.

The excavations have brought to light six inscribed objects of Sargon I: two brick stamps of baked clay, the fragment of a third, and three door sockets. The brick stamps [2] are made from the same mould. The inscription (Pl. 3, No. 3) reads as follows: 1. *Shar-ga-ni-shar-âli* 2. *shar* 3. *A-ga-de^{ki}* 4. *bâni* (BA-GIM) 5. *bît* 6. *^{ilu}Bêl*, "Shargânisharâli, king of Agade, builder [3] of the temple of Bêl." Judging from their appearance, these brick stamps were never practically used, but were presented by Sargon as temple-offerings to Bêl in commemoration of his work; or perhaps they were placed in the corners of the structure erected by him, as was the case with the later clay cylinders. [4] That others which were of the same form as these were used for stamping bricks can neither be proved nor denied. [5]

Of greater importance are the door sockets, which contain the longest inscriptions of Sargon thus far known. Two of these are exactly alike in their contents (Pl. 2). The inscription of the third (Pl. 1) differs somewhat. Pl. 2, as the more important, reads as follows: 1. *^{ilu}Shar-ga-ni-shar-âli* 2. *mâr Itti(-ti)-^{ilu}Bêl* 3. *da-num* 4. *shar* 5. *A-ga-de^{ki}* 6. *û* 7. *su^6-û-la-ti* 8. *^{ilu}Bêl* 9. *bâni* 10. *E-kur^7* 11. *bît ^{ilu}Bêl* 12. *in Nippur^{ki}*, etc.,[8] "Shargânisharâli, son of Itti-Bêl, the mighty king of Agade and of the dominion(?) . . . . of Bêl, builder of Ekur, temple of Bêl in Nippur." From this text we learn the interesting fact that Sargon's father was Itti-Bêl ("With-Bêl"). [9] Inas-

---

[1] *ushziz* never signifies "to restore," but "to set up;" *ênuma lâban*, as Scheil transcribes, could never be (Grammar!?) the Babylonian or even Lulubitic equivalent for "alors qu' elle tombait."

[2] The cuneiform characters have been executed in relief, and are larger at the base than at the top. My copy gives the exact size of the characters at the base, while the photographic reproduction illustrates the size at the top.

[3] *Banû* means to build something or to build at something that already existed, *i. e.*, to add to it or to restore it if it was in ruins. All that we can say of Sargon is that he was a builder of the temple, but not its first builder.

[4] "One of the cylinders from Babylon, now in the British Museum, was not found, as I was able to learn from the man who discovered it, in a corner, but in a niche in the side of a long wall" (Peters).

[5] Winckler's doubts (*Gesch.*, p. 26) are dissipated by the evidence of the phrases *bâni bît Bêl* and *bâni Ekur bît Bêl in Nippur* (Plates 1–3).

[6] Brünnow, *l. c.*, 802 (Jensen). The significance of *sûlati* (or plur. *sûlâti ?*) is not certain. Is כֹּלְלָה (Jer. 33, 4) to be compared ?

[7] This—not *E-shar* (Delitzsch, *Gesch.*, p. 33)—was the name of the temple of Bêl in Nippur. Cf. Jensen, *Kosmologie*, pp. 186 *seq.*, 196 *seq.*

[8] For the rest, cf. pp. 10, 13, 14.

[9] Perhaps shortened from *Itti-Bêl-balâṭu*, "With Bêl is life" (Strassmaier, *Nabon.* 466, 13 ; *Cambys.* 373, 10). Cf. the similar formations *Itti-Marduk* (-*Nabû*, -*Shamash*, -*Gula*, etc.)-*balâṭu* in the Contract literature.

much as the latter does not bear the title of king, we may[1] see therein a confirmation of the legend[2] of Sargon, l. 2, *a-bi ul i-di aḫu abi-ia i-ra-mi sha-da-a*, "my father I know not, whereas the brother of my father inhabits the mountain," viz., that Sargon, being of an inferior birth on his father's side, was a usurper.

My use of *Shargâni-shar-âli* as identical with *Shar-gi-na*—known from the inscriptions of Nabûna'id as the father of Narâm-Sin—requires a word of explanation. Sayce,[3] Hommel[4] and Tiele[5] have never called in question the identity of the two names, reading the name of our king as *Shar-ga-ni*, and regarding *shar âli* as his first title. Similarly Pinches distinguished between the name and the title, at first[6] interpreting the latter with Ménant as *lugal-lag*, "the messenger king," but afterwards[7] with Hommel as *shar âli*, "king of the city." Ménant[8] and Oppert, on the contrary, believe that *Shar-ga-ni-shar-luḫ* (Ménant), or *Shar(Bin)-ga-ni-shar-imsi* (Oppert[9]), or *Shar(Ḫir, Bin)-ga-ni-shar-ali* (Oppert[10]) is to be regarded as one word, containing only the name of the king. More recently Winckler,[11] adopting Oppert's view, reads the name *Shar-ga-ni-shar-maḫâzi*. He considers the identity of this name with Sargon as an open question, whilst Oppert holds it to be simply an *inadmissible plaisanterie*.[12] It is not clear to me what induced Oppert to regard *Shar-ga-ni* as identical with *Bin-ga-ni*.[13] The syllabic value of *bin* for the sign SHAR is unproven, and in itself improbable.[14] On the other hand, I share the view of Oppert-Ménant in

---

[1] This conclusion is very probable, but not absolutely certain, as the title of king is very frequently omitted when the names of the fathers of Cassite kings are referred to, although they are known to have been "kings."

[2] Although evidently containing history interwoven with legend, it is nevertheless historically important, as giving expression to the Babylonian conception of the history of the ancient Sargon. Its value increases in proportion as we find in it statements which are proven from other sources to be correct. Incidentally, it may be remarked that on account of the mention of the father's brother in the "Legend," and because of Sargon's own statement concerning Itti-Bêl, the clause *abi ul idi* can only be regarded as meaning that Sargon did not know his father personally, since the latter was dead (Tiele, *l. c.*, p. 114), or for various reasons was compelled to keep himself in concealment.

[3] Cf. *e. g., R. P².* I, p. 5

[4] *l. c.*, p. 302 *seq.*

[5] *l. c.*, p. 488, note 1.

[6] *P. S. B. A.* VI, pp. 11–13, 68 *seq.* Cf. V, pp. 8, 9, 12; VII, pp. 65–71. *Trans. S. B. A.* VIII. pp. 347–351.

[7] *P. S. B. A.*, VIII, pp. 243 *seq.*

[8] *Recherches sur la Glyptique orientale*, p. 74. *P. S. B. A.* January 5, 1884.

[9] *Collection de Clercq.*, No. 46, p. 50.

[10] *Z. A.* III, p. 124.

[11] *Gesch.*, pp. 39, 327, and Schrader's *K. B.* III, Part 1, p. 101 **. Cf. *Unters.*, p. 44 *seq.*

[12] *Z. A.* III, p. 124. Ibid. : "*quoique roi d'Agade, il n'est pas plus Sargon, que les empéreurs Louis et Lothaire ne sont un même personnage.*" Winckler's article in *Revue d' Assyriologie* II (quoted in *Unters.*, p. 79, note 4), was unfortunately not accessible to me.

[13] In the name *Bi-in-ga-ni-shar-âli* on a seal cylinder, published by Ménant, *Glyptique* I, Pl. I, No. 1. Cf. Winckler, *Altbabylonische Keilschrifttexte* (quoted as *A. K.*), No. 66.

[14] Even if it was proved that *SHAR* has the value of *bin* in a few cases, it would be utterly impossible to give the character this exceptional value in a Semitic word list (V R. 41, l. 29, *a, b*). Cf. p. 18, note 4.

regard to the close connection of these three words as constituting the name of the king, and read accordingly *Shargâni-shar-âli* as one word. For, as Oppert properly states, it is impossible to read the name simply *Shar-ga-ni*, inasmuch as, according to the parallel passages of the oldest Semitic cuneiform texts, in this case we should expect the two parts (*Shargâni* and *shâr-âli*) to be separated by a line. Only individual words, or two expressions very intimately connected,[1] as "son of Itti-Bêl," "temple of Bêl," "in Nippur," are written together without this separating line.[2] Titles are not considered to stand in such close connection with their antecedent proper names.

But, contrary to the view of the two French scholars, I maintain the identity of Ṣargon and *Shargâni-shar-âli* for the following reasons:

1. By the side of the long names of kings and private individuals we find—at least in the last two thousand five hundred years of Babylonian history—abbreviated forms in use. The lists of kings and the contract tablets, not to mention other passages, furnish ample proof. Cf. *e. g. Ki-an* (List b[3]) with *Ki-an-ni-bi* (List a, Rev.); *Kir-gal* (List b) with *Kir-gal-dara-bar; A-dara* (List b) with *A-dara-kalam-ma; Bibe* (List b) with *Bi-be-ia-shû*[4] (Pl. 26, No. 70); *Kab-ti-ia abil-shu sha Tab-ni-e-a,*[5] with *Kabti-ilâni-Marduk abil-shu sha Nabû-tab-ni-u-ṣur,*[6] among hundreds of similar examples.[7] It is therefore highly probable that at some future time we shall find the abbreviated form Shargâni even on Sargon's own monuments.

2. It was especially to be expected in the case of a king famous above all others, and who so early became the hero of popular story, that the longer name should so[8] be abbreviated in the mouth of the people, and, finally, when it had ceased to be intelligible, explained after the method of 'folk etymology',[9] as *Sharru-kênu*, "the true king." Moreover, Pinches[10] has pointed out, by comparison of Sumer. *kurgina* = Assyr. *kurkanû, gishkin* = *kishkanû*, that the sign GI (*ge*) was originally pronounced as *ga*, and that the Hebr. סַרְגּוֹן represents this older pronunciation.[11]

---

[1] In this respect the writer of the *stèle de Zohâb* is freer. Cf., however, *sha duppa*, which is always written on one line even in the Sargon inscriptions from Nippur and in that of the king of Guti.

[2] Cf. Pl. 1, l. 3, 11, 24; Pl. 2, l. 1, 2, 11, 12, 23; Pl. 3, No. 3, l. 1; No. 4, l. 1, 3.

[3] Winckler, *Unters.*, p. 146, col. I, 4. For List a, cf. ibid., p. 145.

[4] Hilprecht, "Die Ergänzung der Namen zweier Kassitenkönige," in *Z. A.* VIII, in print.

[5] Strassmaier, *Nabon.* 133, 4.

[6] Strassmaier, *Nabon.* 132, 4. Cf. Peiser, *Aus dem Babylonischen Rechtsleben* I, p. 11.

[7] The same principle of abbreviating names in everyday use occurs among nearly all ancient nations. Cf. *e. g.*, Erman, *Ägypten und Ägyptisches Leben im Altertum*, p. 233; also the Hebrew dictionaries; Fick, *Die griechischen Personnenamen;* O. Crusius, *Neue Jahrbücher*, 1891, pp. 385–394: "Die Anwendung von Vollnamen und Kurznamen bei derselben Person." For the last two references I am indebted to my friend and colleague, Prof. W. A. Lamberton.

[8] *Shargâni*, "the powerful." See p. 18, note 4.

[9] Hommel, *Gesch.*, p. 301.

[10] *P. S. B. A.*, VII, p. 67 *seq.*

[11] Cf. Hommel, *l. c.*, p. 303.

3. It is absolutely impossible to regard Sargon, father of Narâm-Sin, as "perhaps an invention of legend."[1] But were he one of the best known and mightiest rulers of the olden time,[2] it was to be expected that some monuments of his would be found in the thorough exploration of the ruins of the temple at Nippur, where the greatest number of texts of his time[3] ever found has been brought to light. Where inscriptions of his less known son Narâm-Sin, and of the hitherto altogether unknown Âlusharshid, have been discovered, it was a priori probable that inscriptions of Shargîna = Shargêna = Shargâni(a) would also come to light. Therefore the very absence of the name in the inscriptions there discovered is, in itself, a proof that the ancient king whose name commences with Shargâni, and who is represented by six inscriptions, is no other than Sargon, the father of Narâm-Sin. From this it follows naturally that the later Shargêna was merely an abbreviation of Shargâni-shar-âli.

According to Oppert, the name signifies "mighty is the king of the city."[4]

There were also found in Nippur two brick stamps of Narâm-Sin, son of Sargon I. Both contain the same legend. The moulds, however, that were used in making them differ slightly in size and shape. The inscription reads: 1. ^{ilu}Narâm-^{ilu}Sin 2. bâni 3. bit ^{ilu}Bêl, "Narâm-Sin, builder of the temple of Bêl." If we may base an argument on the place in which the stamps were found, as to the location of Narâm-Sin's building, we might conclude that he built a shrine immediately on the canal south from the Ziqqurratu, whilst his father confined himself in his building to the east side of the temple platform. In any case, from the contents of the

---

[1] Winckler, Gesch., p. 39.

[2] As is proved by the inscriptions of Nabûna'id, where he is called "king of Babylon", by the "Legend of Sargon," the Tablet of Omens IV R. 34, and the mention of his name in the List V R. 44, 18, a, b. Hommel, who reads erroneously Lugal-girinna (l. c., pp. 301, 307, note 4) in the last quoted passage, distinguishes Sargon of the list as Sargon II, c. 2000 B.C., from the ancient Sargon I. His arguments are not convincing (cf. also Winckler, Unters., p. 45, note 2). It is especially "the historical background of the work"—the mention of Elam, Guti, etc., at such an early period, which is the most valuable evidence for the high antiquity and reliability of the statements contained in the astrological work. Cf. my remarks in connection with the inscriptions of the king of Guti and Âlusharshid.

[3] Six inscriptions of Shargâni-shar-âli, two of Narâm-Sin, and sixty-one inscribed vases (or fragments) of Âlusharshid.

[4] Z. A. III, p. 124. Cf. V R. 41, 29 a. b.: shar-ga-nu = dannu. Shargânu is a noun formation in ân (Delitzsch, Gram., § 65, No. 35) from a root sharâgu, which seems to mean "to be powerful, mighty." Cf. the Hebr. proper name שׂרגון. Likewise the names Bingâni-shar-âli and Âl-usharshid contain the formative element âlu. There are reasons for identifying this âlu (Âlu) with Âlu^{ki}, used as an ideogram for "Babylon" by Nebuchadrezzar II (misunderstood by Delitzsch, Wörterbuch, p. 6). Cf. Hilprecht, The Sunday School Times, 1892, No. 20, p. 306 seq. Nebuchadrezzar uses even maḫâzu alone (urbs) for "Babylon." Cf. e. g. V R. 34 (Z. A. II, p. 142–44), col. I, 13: zanân maḫâzi, "to adorn the City" (i. e. Babylon, not "die Städte," Winckler in Schrader's K. B. III, Part 2, p. 39). For the use of Âlu without ki, cf. below Kish (Kishshatu).

inscriptions of Sargon and Narâm-Sin it follows that the dominions of both included Nippur.[1]

The list of ninety-two garments, Pl. 6, was found near the inscriptions of Narâm-Sin. As it is written in Semitic (cf. l. 6, *rabâtum*), and as, paleographically, there is no objection to such a conclusion, it belongs probably to Narâm-Sin, or, in any case, to one of the earliest Semitic kings of Babylonia.

In this connection, I call attention to the interesting and important fact that the fragment of another vase (or probably of several) was discovered in the same deep-lying stratum as the inscriptions of Sargon and Âlusharshid, and close by them. This fragment[2] contains the statement that "*En-te(men)-na*, patesi[3] of Shirpurla," presented the vase to Bêl of Nippur. When to this we add that a vase of Narâm-Sin,[4] and another of Âlusharshid, as I have been informed, was found in Tello, we may safely conclude: 1. That the dominion of Sargon,[5] Narâm-Sin and of their immediate successors (or predecessors[5]) extended also over the whole of South Babylonia[6] (at any rate, as far as Shirpurla[7]). 2. That the chronology of the oldest Semitic rulers of Babylonia is approximately the same[8] as that of the earliest patesis of Shirpurla. 3. That the "kings of Shirpurla" are earlier than Sargon (or Âlusharshid[5]). It was apparently Sargon I or Âlusharshid who put an end to the independence of the kingdom of Shirpurla. This is not the place for a detailed statement of all my reasons. They will be found in full elsewhere.

To the early Semitic rulers of Babylonia already known must now be added, in consequence of the discoveries at Nippur, King URU-MU-USH, as his name is written. Not less than sixty-one fragments of different vases of his have been excavated from the temple.

As to the material of the vases cf. Table of Contents. The fact that they were found close to the monuments of Sargon, that like them they are written in Semitic, that the phraseology of Pl. 4, l. 11, 12 is very similar to lines 6, 7 of the vase inscrip-

---

[1] Cf. above, p. 15, note 5, and p. 25, note 3.

[2] It will be published in Vol. I, Part 2.

[3] I hold that the change of the title of *lugal* into *patesi* in the case of the princes of Shirpurla is an indication of their political dependence (Hommel, *l. c.*, p. 296). Jensen's view (Schrader's *K. B.* III, Part 1, pp. 6–8) is somewhat different.

[4] According to Oppert. Cf. Hommel, *Gesch.*, pp. 299, note 1, 309.

[5] See my remarks in connection with the texts of Âlusharshid.

[6] Cf. Hommel, *l. c.*, pp. 296, 311.

[7] Winckler's suggestion that Shirpurla is not identical with the modern Tello or part of these ruins (*Gesch.*, pp. 24, 31, note 1, 44, 326), but that it lay in North Babylonia, is quite improbable, to me even impossible.

[8] In this I slightly differ from Hommel (*l. c.*, p. 296), who places Sargon and Narâm-Sin a little later than the oldest patesis of Shirpurla.

tion of Narâm-Sin, that paleographically they show the characteristic features of the inscriptions of Sargon and his son, all this points to the first half of the fourth millennium as the approximate date when they were written. As the language of the inscriptions is Semitic, I regard the name of the king also as Semitic and read tentatively *Âlu-usharshid*,[1] *i. e.*, "He (some deity) founded the city."[2]

The discovered inscriptions of this king may be classed in four groups, consisting of thirteen, eleven, six and three lines respectively. Only three of the three-line legends[3] have been preserved intact. Though not a single complete text of the six-line inscriptions has been excavated, yet the faint traces to be seen in the third-line of Pl. IV, No. 13, and the space left for the restoration of the text, justify my reading of Pl. 5, No. 6, l. 1–3. The fragment reproduced on Pl. 5, No. 10, is the only remnant of an eleven-line inscription found at Nippur. It is in all respects similar to the thirteen-line inscriptions, with this difference only that l. 11, 12 of the latter, *in namrak Elamti*[ki], were omitted. The inscription of thirteen lines has been reconstructed from eleven fragments, three of which (Pl. III, Fragm. 8891, 8892, a, b) belonged to a large dolomite vase and formed the basis of my text. Eighteen fragments of all the excavated vases may confidently[4] be referred to this group. The long inscription, of which some of the shorter ones are possibly abbreviations,[5] reads:
1. *A-na*   2. [ilu]*Bêl*   3. *Âlu-usharshid*   4. *shar*   5. *Kishshatu*   6. *i-nu*   7. *Elamtu*[ki]
8. *ù*   9. *Ba-ra-'-se*[ki]   10. *inîra*   11. *in nam-ra-ak*[6]   12. *Elamti*[ki]   13. *iddin* (A-MU-

---

[1] Cf. Brünnow, *l. c.*, 5032, 5068.

[2] Cf. Hilprecht, *Z. A.* VII, p. 315, note 1, and Pinches, *The Academy*, September 5, 1891, p. 199. Even if the name be transliterated *Urumush*, it may be Semitic. In this case the *Orchamus* of Ovid (*Metam.*, 4, 212) offers itself for comparison.

[3] In spite of their identical contents I reproduced two of them (Pl. 5, Nos. 7 and 8), because of the slight difference in the form of the characters USH and *sharru*, and because we do not possess a superabundant supply of texts dating from that ancient period to which they belong. The sign published on Pl. 5, No. 9, and resembling the Old Babylonian character for *ilu*, "god," is found on the bottom of a third vase of the three-line group, and is, no doubt, merely a "trade-mark."

[4] I include here only those fragments of which portions of l. 5–13 have been preserved. Some of the other fragments, however, probably belong to the same group.

[5] Necessary because of limited space.

[6] This word has been variously translated. Tiele (*Gesch.*, p. 115) and others before and since changed *namrak* into *Apirak*, a city mentioned on the tablet of omens, col. II, 12–14. Hommel (*Gesch.*, pp. 279, 309) translates it "polished work," whilst Winckler (*Gesch.*, p. 38) is content to render it simply "work." But all this is mere guess work. To my knowledge, the word has been found thus far only in three passages, in the above text of Âlusharshid, on the vase of Narâm-Sin and in Gudea B, col. 6, 66. In the last passage we read l. 64–69 : *gish* KU [uru]*An-sha-an Nima*[ki] *mu-sig nam-ra-aga-bi* [dingir]*Nin-gir-su-ra E-ninnû-a mu-na-ni-tur*, "With (his) weapon he smote the city of Anshan in Elam, brought its spoil into Eninnû to Ningirsu." Cf. Jensen (*K. B.* III, Part 1, pp. 38, 39) on this passage. The latter's hesitation about the reading *Nima*[ki], "Elam" (exactly so written above), and the meaning of *namrak* is unnecessary. As early as eight years ago, Amiaud, with his wonted insight, conceived the correct meaning of the word (*Z. K.* I, p. 249). Whether it is Sumerian or Semitic remains to be determined. As we do not possess long

SHUB),[1] "Âlusharshid, king of Kishshatu, presented (it) to Bêl from the spoil of Elam, when he had subjugated Elam and Bara'se."

The inscription is of historical importance. We learn from it, that King Âlusharshid subdued Elam and the country of Bara'se, doubtless in close proximity to it,[2] and that in the booty he carried off to Babylonia a number of costly marble vases. Part of them he dedicated to Bêl of Nippur, and part, perhaps, to Shamash of Sippara,[3] after first having engraved upon most[4] of them in beautiful clear-cut characters his name and the occasion of the gift. The inscription suffices to show that Âlusharshid was a mighty ruler, who in courage and adventurous spirit was not second to Narâm-Sin. But it also offers most welcome material for determining the extent of the dominion of the oldest Semitic rulers. It furnishes additional support to Tiele's view (*Gesch.*, p. 114), and at the same time proves that Winckler's conception of the beginning of the North Babylonian history and of the extent of Sargon's empire (*Gesch.*, p. 38) is incorrect. Winckler proceeds upon the erroneous supposition that the deeds of Sargon, as reported in the tablet of omens and in the "legend," are purely legendary. Hommel also (*Gesch.*, p. 306 *seq.*) is hampered by similar prejudices. That Narâm-Sin was in the possession of South Babylonia is demonstrated by his building in Nippur (*bâni bît Bêl*), and by his vase found in Tello, and is furthermore established beyond all doubt by his successful operations in Magan,[5] which, according to Winckler, was situated on the eastern boundary of Arabia. A vase of the Semitic king of Guti,[6] belonging to this same ancient period, which was probably carried by a victorious Babylonian king as trophy to Sippara, points to the extension of the power of the oldest North Babylonian rulers

descriptions of campaigns in Sumerian, it cannot be surprising that the word does not occur otherwise in Sumerian inscriptions, which deal mostly with religious affairs and accounts of buildings. In favor of a Semitic etymology, to which I incline, it may be said : (1) That the word "looks very much like an original *m*-formation of a root מרך" (Jensen) and (2) that it is twice found in the Semitic inscriptions of the oldest North Babylonian rulers.

[1] It is not to be read *a-mu-ru* and to be derived from *amâru* with the meaning of "*ersehen*" (Hommel, *Gesch.*, p. 302), *i. e.*, "to dedicate" (Pinches, *Trans. S. B. A.* VIII, p. 350). Cf. Amiaud, *Z. A.* II, p. 296, and Jensen in Schrader's *K. B.* III, Part 1, p. 26, note *⁰. For *shub = nadânu = nadû* (נרה, cf. גֶרֶה, "gift," Ezek. xvi. 33), cf. Tallquist, *Babylonische Schenkungsbriefe*, p. 9.

[2] Nothing more definite can be said at present. It is, perhaps, to be read *Para'se*. Cf. the name of the mountain *Ba-ti-ir* (*stèle de Zohâb* I, col. I, 7), which Scheil (*l. c.*, p. 104) correctly identified with the mountain *Pad(d)ir* (*Shamshi-Rammân* II, col. II, 7).

[3] According to Pinches Jensen, inscriptions of Âlusharshid have also been found in Sippara. Cf. *The Academy*, September 5, 1891, p. 199, P. S.

[4] A number of vases of the same high workmanship and found among them were without inscriptions. Cf. below, p. 80.

[5] I. R. 3, No. VII, l. 7, *namrak Magan*, "plunder of Magan."

[6] Cf. p. 12 *seq.*

further northward. The inscriptions of Âlusharshid testify to his supremacy over the South,[1] and to his victories in the East and North-East of Babylonia. In view of all this, I regard it as impossible to question the historical character of the statements of the tablet of omens relative to Narâm-Sin. Since we know that about that time a Semitic population dwelt in the northern and northeastern countries of Guti and Lulubi,[2] whose kings wrote inscriptions on rocks and vases in a dialect entirely identical with the Babylonian, it can no longer seem strange that Narâm-Sin took the Semitic king *Rish-Rammân*, of Apirak, prisoner. It is evident, however, that Apirak, which by its termination forcibly recalls names like A(E)shnunak,[3] is to be sought in the North-East[4] of Babylonia rather than in the South.[5] If the credibility of the tablet of omens is therefore established as far as Narâm-Sin is concerned, we are no longer at liberty to call in question what it relates concerning Sargon I, unless more solid objections than have heretofore been raised, be brought against it. With Tiele, therefore, I regard as facts what Winckler describes as fiction, viz., that Sargon I subjugated nearly the whole world known to him, or in other words, "the four quarters of the earth." [6]

But how is it that whilst Sargon always bears the title *sharru dannu shar Agade* or *dannu shar Agade* or only *shar Agade*,[7] both in the legend and in his own inscrip-

---

[1] Including Lagash. Cf. p. 19.

[2] This fact argues in favor of a migration of the Semites into Babylonia from the North. Cf. the "legend of Sargon," according to which his uncle dwelt in the mountains, and he himself was carried down the river in an ark made of reed. Cf. also Winckler, *Gesch.*, p. 141.

[3] Pognon found there Semitic inscriptions written by *patesis* of Ashnunak. Nothing can be said with certainty as to the exact date of these texts, but they seem to belong to the second millennium B. C. Cf. Pognon, *Quelques rois du pays d'Achnounnak*, read at the *Académie des inscriptions et belles lettres*, March 18, 1892. On this country see further Delitzsch, *Paradies*, p. 230 *seq.*; *Kossäer*, p. 60 ; and also Jensen in Schrader's *K. B.*, Part I, p. 137, note⁰.

[4] Hommel is on the right track (*Gesch.*, p. 310, note 1). His reading *A-ma-rak*, however, has neither support nor probability.

[5] Delitzsch, *Paradies*, p. 231, "*ziemlich südlich zu suchen.*"

[6] I regard also Sargon's campaign in the West, to the Mediterranean Sea and to Cyprus, as historic facts. The cylinder of Narâm-Sin's servant found at Cyprus, and now in the Metropolitan Museum of New York (cf. Sayce, *Trans. S. B. A.* V, p. 441 *seq.*), has, however, no direct bearing upon the whole question. Through the kindness of Prof. Isaac Hall, Curator of the Museum, I obtained an accurate impression of the cylinder, to which, for paleographic reasons (observe, *e. g.*, the form of the character *ra*), I cannot assign an earlier date than c. 2000–1500 B. C. The pictures on it also point to a more recent date. But the cylinder is undoubtedly no modern forgery (Hommel, *l. c.*, p. 309).

[7] Nabûna'id calls him, for apparent reasons, *shar Bâbili*. It is in itself not impossible that there were kings of Babylon at some time in that ancient period. For the place where the vase of Narâm-Sin was found by the French expedition, the tablet of omens (I, 7–11, cf. my restoration of this passage below, p. 26) and the occasional mentioning of Babylon (under another name) in the Sumerian inscriptions of the kings and patesis of Shirpurla clearly show that Babylon not only existed at this early time and belonged to Sargon's kingdom, but that it even had already obtained considerable prominence (cf. below, p. 26). Cf. however, Winckler, *Unters.*, p. 76 *seq.*, and Lehmann, *Shamashshumukin*, p. 96, note 4.

tions, his immediate successor, Narâm-Sin, styles himself *shar kibrat arba'i*, and Âlusharshid and MA-AN-ISH-TU-SU[1] even *shar Kishshatu?* This question is closely connected with the other, What do the last two titles mean? It is impossible for me to enter here into as full a discussion of this question as its importance demands. I therefore content myself for the present with giving the results of my investigations. As I am now considering the meaning of these titles in the earliest times only, I naturally exclude their use with the later Babylonian and with the Assyrian kings.[2]

I. As to the Old Babylonian title, *shar Kishshatu*, we have been accustomed to follow Winckler,[3] and to regard it as simply the equivalent of the later *shar kishshati*, "king of the world."[4] This identification, however, is not proved. On the other hand, it is worthy of note, (1) that supposing Âlusharshid lived after Narâm-Sin, and even supposing further that he founded a new dynasty, it would still be matter for astonishment that he should exchange a title, that was not only satisfactory to Narâm-Sin, known as a great conqueror, but was in itself sufficiently significant, for the synonymous *shar kishshati*, "king of the world;"[5] (2) that no later Babylonian king, before Merodachbaladan I, not even the powerful Hammurabi, bears this title, though many of them apply to themselves the title *shar kibrat arba'i*; (3) that Winckler's theory, which sees in Harran the original seat of the *sharrût kishshati*, is improbable for the later Babylono-Assyrian time, and altogether out of question for

[1] Winckler, *A. K.*, No. 67. Paleographic reasons, the Semitic language of the inscription and the title *shar Kishshatu*, establish for this king a date not only earlier than 2000 B. C. (Winckler, *Gesch.*, p. 155), but even earlier than 3000 B. C. He is to be classed with Âlusharshid. The white marble duck (Norris, *On the Assyrian and Babylonian Weights*, Pl. 2, No. 2), bearing the name of *Nabû-shum-lîbur shar Kishshatu*, remains without consideration here, as I do not feel at liberty to base any paleographic conclusions on the cuneiform text as it is published there.

[2] I hope to treat the whole question in another place. That we may understand correctly the meaning of this title in Assyrian, the following points must be examined more carefully : (1) Is the title simply to be regarded as borrowed from Babylonia (cf. *patesi*, temple names, etc.) and extended to cover Assyrian conditions, so that only the name is Babylonian, while its semasiological development is essentially Assyrian? (2) Or, in using the title, did the Assyrians claim the same right over the same district as the Babylonians, *i. e.*, suppose that in Babylonia a claim was thereby expressed to Harran (Winckler), did the Assyrians by their use of the phrase make exactly the same claim upon this city? (3) Or is there no connection between the Assyrian and the Babylonian title? These questions have hitherto not been answered sufficiently.

[3] *Mitteilungen des Akademisch-Orientalischen Vereins zu Berlin* I, p. 14.

[4] Cf. Jensen in Schrader's *K. B.* III, Part 1, p. 196, note 4.

[5] If we may draw any conclusion from the later customs of Babylonian and Assyrian kings, we rather expect that in the above given case, Âlusharshid, whose empire was scarcely smaller than that of Narâm-Sin, according to our present knowledge, would have been particularly anxious to adhere to a title which was connected by the Babylonian people with the name of a very powerful ruler, and regarded by the later kings as especially important. And *vice versa*, if Âlusharshid lived before Sargon and had founded a *sharrût kishshati*, "kingdom of the world," it would be strange that Narâm-Sin should have used *shar kibrat arba'i* instead, if the other title meant exactly the same.

the earliest period.[1] I therefore would propose another explanation of the title, viz., to regard *shar Kishshatu* (or *shar Kish*) as identical with *shar Kish*, "king of Kish."[2] In other words, I infer from this title that there was a kingdom of the city of Kish similar to those of Shirpurla, Agade, etc., at the earliest time of the Babylonian history. Two of its rulers are so far known; both wrote Semitic, and one of them at least possessed South Babylonia and defeated Elam. Whether these kings lived after the dynasty of Sargon, or whether they preceded it and were dethroned by Sargon, will be considered below. At all events, it will be well to separate the kings of Kish[3] from those of Agade. There is much in favor of the view that even in the Assyrian mind[4] the title *shar kishshati* was originally connected with the possession of Kish, where Tiglath-Pileser III offered sacrifices to the gods (II R. 67, 11).

II. But what does *shar kibrat arba'i* mean in the oldest Babylonian history? After Sargon had subjugated the Elamites,[5] thus fixing the natural eastern boundary of his projected great empire, he marched to the West, "subdued 'the land of the West,' conquered the four quarters of the world." The last part of the previous sentence, literally quoted from the tablet of omens, can in itself be interpreted as meaning (*a*) that "the four quarters of the world" lay still beyond "the land of the West," and therefore were geographically distinct from it, or (*b*) that the conquest

---

[1] Cf. also A. Mez, *Geschichte der Stadt Harrân in Mesopotamien*, p. 27.

[2] As I remarked above, I cannot state all the reasons for my theory here. At present it may suffice to give the following : (1) Cf. my restoration of IV R. 34, 7–11 below. (2) Cf. Delitzsch, *Paradies*, p.218 *seq.*, where it is stated that the Semitic Babylonians and Assyrians wrote this city also *Ki-shu* (and *Ki-e-ish*, Brit. Mus., 82–8–16, 1, col. I, 44, published by S. A. Smith, *Miscellaneous Assyrian Texts*, Pl. 26 ; cf. also the present volume, Pl. 8, No. 14, l. 7), and *Kish-sha-tu*, "according to a small unpublished vocabulary" (cf. *Paradies*, p. 230). (3) Cf. also the name of the ancient king, *Abil-Kish ki*, known from the fragment of a Babylonian chronicle (*Trans. S. B. A.* III, 372), and to whom Delitzsch (*Gesch.*, p. 72) correctly assigns the fourth millennium.

[3] I afterwards found that Jensen (*Schrader's K. B.* III, Part 1, p. 202, note), independently of me, translated "king of Kish" in the inscription of Manishtusu (Winckler, *A. K.*, No. 67). His reasons for so doing and his conclusions are both unknown to me.

[4] The facts that Rammàn-niràri, who defeated the Babylonian king, Nazi-Maruttash, near Kâr-Ishtar, is the first Assyrian ruler who bears the title *shar kishshati* (in the inscription of his son, Shalmaneser I, I R. 6, No. IV, l. 2); and further, that Tukulti-Ninib I, his grandson, who also claims the title, must have been in the possession of Kish, as he had captured even Babylon (*R. P.²*, Vol. V, p. 111, col. IV, 2 *seq.*); and last, that neither Ashurdan I, nor Mutakkil-Nusku, nor even Ashur-rêsh-ishi has this title (III R. 3, No. 6, l. 1 and 8), deserve especial attention in connection with my hypothesis. Afterwards the ancient meaning of the title was lost, and *shar Kishshati*, "king of Kish," became *shar kishshati*, "king of the world" (which may, however, have been the very first meaning of the title before it was connected with Kish; cf. the development of the meaning *shar kibrat arba'i*).

[5] IV R.² 34, col I, 1–3. I regard the arrangement of the individual deeds, related in the tablet of omens, as chronological. Among other reasons the account of Sargon's three expeditions against the West favors this view. It was also natural that the king, before marching to the West, should protect himself in the rear by subjugating the Elamites in the East, so that during his long absence no danger might threaten Babylonia from that quarter.

of "the four quarters of the world" is identical with his conquest of "the land of the West," or (c) that the conquest of "the four quarters of the world" followed as a result upon his subduing the West. In opposition to the first view is the fact that a kingdom of "the four quarters of the world" in the far West is nowhere else mentioned, that the phrase stands without the usual determinative *mâtu, âlu,* etc., and that this title was claimed by Babylonian kings even when they made no conquests in the West.[1] The identification of the "four quarters of the world" with "the land of the West" needs no refutation, as it has never been advanced, and in fact has no support. We can, therefore, only regard the conquest of "the four quarters of the world" as the result of Sargon's victories in the West, so that by the use of the title the claim is made to a quasi-worldwide dominion,[2] as has been correctly stated by Lehmann (*l. c.,* p. 94). And indeed, Sargon, after having conquered the West, was fully justified in the Babylonian sense of the word "world," in thus designating his large dominion. For, in order to subjugate the West, he was obliged, because of the Arabian desert, to march victoriously first to the North, then to the West and finally southward. The enemies in the East having been previously subdued, and South Babylonia being also brought under his sceptre,[3] he could indeed call a kingdom his own which was enclosed on all sides by natural boundaries.[4]

The city which had obtained the hegemony through Sargon's deeds was Agade.[5] For he calls it "my city" ("Legend," l. 26). It is the city in which he was shut up during the insurrection against him (IV R.[2], 34, col. I, 37). And furthermore, in all his inscriptions as yet found, he calls himself "king of Agade." But, if I understand the tablet of omens correctly, Agade does not appear to have been the capital of the empire of the four quarters of the world, as one would naturally have supposed. After Sargon had subjugated "the whole world," he regarded as his next work the building of a capital worthy of this grand empire. The account of this important work is evidently related in IV R.[2], 34, l. 7–10, a passage[6] unfortunately much mutilated and heretofore entirely misunderstood. After a careful comparison

---

[1] Against Tiele, *Gesch.,* p. 78.

[2] Tiele (*l. c.,* pp. 73, 78) concedes the possibility, indeed even the probability of this explanation, but adds, that the title may also have had an entirely different meaning (p. 73). But what else could it have meant with Sargon I?

[3] This is evident from his building in Nippur, and from the fact that even his son, who was less prominent than his father, extended his influence to Shirpurla. Cf. also the express statements of the "Legend."

[4] The Elamite mountains on the east, the mountains of Armenia on the north, the Mediterranean Sea (and Cyprus) on the west and the Persian Gulf on the south.

[5] In spite of all that has been said in support of *Agane,* I regard this reading as improbable (cf. my remarks on *Gande,* p. 28). Lehmann's statements (*l. c.,* p. 73) prove nothing against *Agade.* More as to this in another place.

[6] For recent translations cf. Hommel, *Gesch.,* p. 305, and Winckler in Schrader's *K. B.* III, Part 1, p. 102 *seq.*

of the text as given in the first and second editions of IV R.,[1] I transliterate and restore the passage as follows: *Shar-ge-na sha ina SHIR an-ni-i Kish-shu* [[ki2]] *Bâbilu* [ki] *i-[shú-]*[3] *shum-ma eprê sha* [(4] *shal-la bâbu TU-NA* [4)] *is-su-ḫu-ma* . . . . *[ina lime?]-tu A-ga-de* [ki] *âlu i-bu-shu-ma* [UB-DA][5] *-*[ki] *shum-shú im-bu-u* . . . . *[ina lib-] bi u-she-shi-bu,* "Sargon, who under this omen brought sorrow upon Kîsh and Babylon, tore away the earth of . . . . and built a city in the vicinity of (or "after the pattern of"?) Agade, called its name 'place (city) of the world,' and caused the inhabitants of Kîsh and Babylon (?) to dwell there."

I infer from this (a) that Kîsh and Babylon existed as prominent cities already in the time of Sargon I, as this great ruler deemed it necessary to render them harmless; (b) that the dynasty of Kîsh was overthrown by Sargon I,[6] and that therefore Âlusharshid and Manishtusu are to be placed before Sargon I;[7] (c) that the reason why the vases of Âlusharshid, all badly broken, were found lying close by the comparatively well-preserved monuments of Sargon, but not by those of Narâm-Sin, is that Âlusharshid apparently ruled before Sargon, not after Narâm-Sin.

The question arises, Which city corresponds in later times to that built by Sargon "in the vicinity (?) of Agade," and with which the title "king of the four quarters of the world"[8] was associated? There are reasons for identifying it with Kutha, as Winckler[9] does. But stronger arguments seem to point to Ursagkalama[10] with its famous temple, "the mountain of the world," (always mentioned in close connection with Kîsh, the probable seat of the *sharrât kishshati*), as being identical with "the city of the world"[11] founded by Sargon I.

[1] This important text seems to have suffered still more since its first publication by George Smith in IV R.[1], as a comparison with Pinches' new edition clearly shows. Had all the differences between the first and second editions of the text, brought about by a decomposition of the tablet, been carefully noted, it would have been of great value, as the first edition is not always accessible to students.

[2] Cf. V R. 12, No. 6, 50 ; II R. 52, 67 c : *Ki-shu* (cf. above, p. 24, note 2). Perhaps *ki* is wanting, and *u,* "and," is to be substituted.

[3] This is the most probable reading, according to the traces in IV R.[2]. Cf. K. 3657, col. I, 9 (*i-shú-ush*), and IV R.[2] 1, * 42, a, "the sickness which brings woe upon the country" (*i-ash-sha-shú*).

[4] These five characters are not quite clear to me, though it is evident that Sargon purposely destroyed something.

[5] The two wedges beginning the character UB are clearly to be seen in IV R.[1], and the last two wedges of DA still remain in IV R.[2]. More than two characters cannot have stood there. For the meaning of *UB-DA,* without *arba'i,* cf. Jensen, *Kosmologie,* p. 167.

[6] For various other reasons the city kingdom of Kîsh cannot be placed after Sargon I.

[7] Paleographical reasons also favor this chronological arrangement of the two dynasties. I reached my conclusion after the plates in question were printed. Pl. 4–5 and III–V are to be placed before those of Sargon I and Narâm-Sin.

[8] It is quite possible that monuments of Sargon may yet be found, on which he calls himself "king of the four quarters of the earth."

[9] *e. g., Gesch.,* pp. 31, 33.

[10] For this reading cf. Jensen in Schrader's *K. B.* III, Part 1, p. 22, note 5.

[11] Cf. Winckler's remarks, *l. c.,* p. 33, in connection with "*Charsagkalama.*"

THE DYNASTY OF ISIN.[1]

Three kings of this dynasty were among the builders of the temple at Nippur, *Ur-Ninib, Bur-Sin I*, and *Ishme-Dagân*.[2] Specimens of brick legends of the latter will be given in the second half of this volume. The fragment of a stone published on Pl. 9, No. 17, is unfortunately so small that we learn nothing new from it.

More important are the inscriptions of both the other rulers, Pl. 10 and 11. They are taken from bricks which, at the time of their excavation, were out of their original place. These formed rather part of a platform of the Ziqqurratu constructed or restored by Mili-Shikhu, who took them from the ruined walls of his predecessors, as old but still serviceable material for his own work. Various bricks of Ur-Ninib have thus been preserved, all with the same inscribed (not stamped) legend. Of Bur-Sin, on the other hand, only a single brick, broken in two pieces, has as yet been found.

Ur-Ninib, " Man (servant) of God Ninib," is the king hitherto wrongly transcribed as Gamil-Ninib.[3] His inscription, here published, is identical with IV R.[2] 35, No. 5. The fragment of a brick from Nippur, I R. 5, No. XXIV, erroneously ascribed to Ishme-Dagân, is obviously the lower half of the same legend. In addition to the complete name of the ruler, the new text offers the correct reading of l. 4, *na-gid*,[4] *i. e.*, *nâkidu*, Hebr. נֹקֵד, "shepherd" (of Ur), and of l. 6, *mí-shú-íl*, "he who delivers the commands" (of Eridu).

Bur-Sin I, so designated by me to distinguish him from another king of the same name,[5] Bur-Sin II of the second dynasty of Ur,[6] is a new king of the dynasty of Isin. The phraseology of his inscription is very similar to that of Ur-Ninib and Libit-Anunit[7] (I R. 5, No. XVIII), and thereby assures the correct reading of several characters of the latter inscription. The first sign of l. 4 is not *da* (Winckler) but *ingar*[8] (identical with Brünnow, *l. c.* 1024), and the second sign in l. 8 is probably

---

[1] Not *Nisin*, as has been generally read—last by Delitzsch, *Geschichte Babyloniens und Assyriens*, p. 79. Cf. the hymn 80, 7-19, 126, l. 3, 4, published by Bezold in *Z. A.* IV, p. 430.

[2] Pl. 9, No. 17, has been placed before Plates 10 and 11 only to save space. Ishme-Dagân was the last king of the dynasty of Isin.

[3] Cf. Hilprecht in *Z. A.* VII, p. 315, note 1.

[4] For this Semitic loan word of the Sumerian language, found also in the inscriptions of Gudea (F. col. IV, 12), cf. Jensen-Zimmern in *Z. A.* III, 200, 208 *seq.* Cf. also Jensen in *K. B.* III, Part 1, p. 4.

[5] Although always written with the other sign *Bur* (Brünnow, *l. c.*, 9068).

[6] Cf. Plates 12, 13, and Vol. I, Part 2.

[7] According to Winckler in Schrader's *K. B.* III, Part 1, p. 86, *Libit-Ishtar*.

[8] Cf. Jensen-Zimmern, *Z. A.* III, p. 199 *seq.*

mí, not *ash*. L. 3–7 in the inscription of Bur-Sin I are of special interest. They read: 3. *íngar lig*(?)[1]-*ga* 4. *Urum*[ki]-*ma* 5. *gish-kin Urudug*[ki]-*ga ki-bi-gi* 6. *ín mí-a-tum-ma*[2] 7. *Uru*[ki]-*ga*, "the powerful shepherd[3] of Ur, the restorer of the oracle tree[4] of Eridu, the lord who delivers the commands of Erech."

GANDE AND THE CASSITE DYNASTY.

A number of inscribed objects excavated in Nippur bear the name of a king[5] who has been transliterated *Gar-de* (?) by Pinches.[6] As I remarked in another place,[7] this transliteration is incorrect. For the first character of the group on Pl. 14, No. 23, l. 2 b, is not the Old Babylonian sign for GAR,[8] but GAN.[9] The second character may be read either *de* or *ne*, the whole name therefore, either *Gande* or *Ganne*. The former reading is the more probable, because the second character, outside of the purely Sumerian[10] texts, is more frequently found with the syllabic value *de* than *ne*.[11]

The contents of the three inscriptions of Gande published on Pl. 14 are identical. They read: 1. [Dingir]*En-lil-la* 2. *lugal ki-aga-ni Gan-de* 3. *a-mu-na-shub*, "To

---

[1] Cf. Jensen, *Z. A.* I, p. 396, note 4.

[2] *mi-a-tum-ma*, corresponding to *mi-shú-il* (Ur-Ninib, l. 6), as *tum*, like *il*, is explained by *abâlu*, "to bring, to deliver." Cf. *IV R.*[2] 35, No. 6, 12, 13.

[3] Cf. *ik-ka-ri Ba-bi-i-lu* " (Nebuchadrezzar II), shepherd of Babylon " (Abel-Winckler, *Keilschrifttexte*, p. 33, l. 19). *Íngar = ikkaru*, Hebrew אִכָּר, is a Semitic word adopted by the Sumerian language (Zimmern, *Babylonische Busspsalmen*, p. 5, note 1), and means "farmer," *Landmann* (Jensen-Zimmern, in *Z. A.* III, p. 199 *seq.* ; Delitzsch, *Assyrisches Wörterbuch*, pp. 400–402). In view of the principal occupations of the farmer—tilling of the ground and stock-raising—the word occurs as a synonym either of *irrishu*, talm. אֲרִיסָא (*Z. A.* III, p. 200), or of *nâkidu*, *rîd alpi* (*Z. A.*, ibid.). Accordingly, it is to be translated either as "farmer" or as "shepherd." The latter meaning is the only possible one in the above-given passage, as the context and a comparison with Ur-Ninib, l. 4—*na gid Urum*[ki]-*ma*, "shepherd of Ur'—clearly show. The same meaning is also to be preferred to *Landmann* (Jensen, in Schrader's *K. B.* III, Part 1, p. 59) in passages like Gudea F, col. III, l. 14, where *íngar* stands parallel with *utul*, *sib* and *nagid*, all words for "shepherd."

[4] Cf. Jensen, *Kosmologie*, pp. 99 *seq.*, 249, note.

[5] That the bearer of this name was a king is certain (against Pinches), notwithstanding the omission of the title. Cf. Hilprecht, "Die Ergänzung der Namen zweier Kassitenkönige," *Z. A.* VIII (in print).

[6] *The Academy*, 1891, September 5, p. 199, *a, b.*

[7] *Z. A.* VII, p. 315, note 1.

[8] Amiaud et Méchineau, *l. c.*, No. 105.

[9] Ibidem, No. 79, sign 5.

[10] To be understood in the sense established by Lehmann, *Shamashshumukin*, pp. 62–108.

[11] For this and other reasons I reject the reading *Agane* instead of Agade (= *Akkad!* in spite of Lehmann, *Shamashshumukin*, p. 73). Cf. also Hommel, *Gesch.*, p. 302.

Bêl, his beloved lord, Gande has presented it." But who was this Gande who left his name on a number of marble vases,[1] on a large unhewn block of white marble, on two others of reddish granite and on the edge of two door sockets belonging to former Babylonian kings? A due consideration of the following points will enable us to answer the question.

1. The short inscription of Gande just translated is written not only on his own monuments by this king, but is also found on the rough edges of a door socket of Sargon I, and another of Bur-Sin II. Hence it follows, that Gande must have lived after their time, *i. e.*, after c. 2400 B. C.

2. On the other hand, it follows from the depth of the place in which the stones were found and also from the peculiar characters of the inscriptions (see below), that Gande could not have ruled after Mili-Shikhu, or, as the immediate seven or eight predecessors of the latter are known, not after c. 1240 B. C.

3. It is remarkable that Gande by two of his inscriptions characterizes door sockets which had previously been presented to the temple as his own gifts. It is in itself clear that these inscriptions cannot be regarded in the sense of inventory labels, as they are sometimes found in connection with Egyptian antiquities. Only one explanation seems possible, namely, that Gande was not a native king, but invaded and conquered Babylonia and regarded the property of the temple in Nippur as his legitimate spoil. As however he, with his victorious hordes, did not leave the subjected country again, but usurped the Babylonian throne, thereby becoming the founder of a new dynasty, the conquered cities and temples became part of his new empire, to which he now restored the trophies of his victory as his own personal gifts. Had he left Babylonia, he certainly would have carried away the treasures of the temple as spoil to his own country, just as Âlusharshid and Narâm-Sin did, after they had conquered Elam and Magan, or Nebuchadrezzar I, after the destruction of Jerusalem.

4. This explanation of Gande is supported by the character of his inscribed objects and by the peculiarity of their cuneiform writing. All his inscriptions are carelessly executed and are engraved very shallowly; indeed, those on the door sockets and large blocks are only scratched in the unhewn stone. Besides, the characters employed violate the laws which underlie the regular development of the Babylonian cuneiform writing. They appear to have been cut by men unaccustomed to use the chisel in writing, who, it is plain, had adopted the Babylonian system of writing, even endeavoring to imitate the characters of a certain period,[2] but who were neither familiar with their original meaning, nor with the

---

[1] Cf. Vol. I, Part 2.

[2] Cf. *e. g* the characters of the inscriptions of Ur-Nina, de Sarzec, *Découvertes*, Pl. 31, No. 1.

exact form then in use. The scribe regarded *e. g.* GAN (Pl. 14, No. 23) as the doubled form of a certain sign resembling the reversed ancient SAG.[1] For occasionally he divides this character into halves, placing one after the other (Pl. 14, No. 24, 25). The artistic execution of the vases themselves stands in striking contrast to the rude appearance of the inscriptions on them and on the large stones. As a number of uninscribed vases of similar form and of the same skillful workmanship were found together with those of Âlusharshid, there is every reason to believe that Gande's vases formed originally part of the former's gift to the temple, the more so as they were found in close proximity to those of that very ancient king. Only the unhewn blocks of marble and granite, apparently intended for door sockets, were genuine gifts of Gande, probably brought from the Elamite mountains. From the fact that the place occupied by the inscription was not polished or even smoothed, we likewise infer that the scribes of this ruler had neither the artistic taste nor technical training of the Babylonian stonecutters.

5. The name *Gande* has not a Babylonian sound. Besides, it is sometimes found abbreviated into *Gan*. This peculiarity of abbreviating names is characteristic of the rulers of the second and third dynasties of Babylon, as is shown by comparing List b with List a and with the inscriptions of Bibeiashu.[2] Only one king fulfills the requirements (viz., a foreigner, founder of a new dynasty, a prince whose name begins with *Gan*, and who lived between c. 2400 and c. 1240 B. C.). This is Gandash, the first ruler of the Cassite dynasty, which occupied the throne of Babylonia for five hundred and seventy-six years. Gande (otherw. Gan) is abbreviated from Gandash[3] in the same way as Bibe from Bibeiashu.[4]

It is significant that, with the exception of fragment Brit. Mus. 84–2–11, 178 (see note 3), no monument of the founder of the Cassite dynasty and very few of its other members have, up to the present, been found outside of Nippur. This latter was, as I shall later show in detail, the very centre and stronghold of the Cassite dynasty. It is not, therefore, accidental, that the representatives of this foreign house dedicated so many valuable gifts to the temple of Bêl in Nippur. By not paying the same homage to Marduk of Babylon and his illustrious city, which Hammurabi[5] had endeavored to raise to the most prominent position in the political and religious life of the country,

---

[1] Amiaud et Méchineau, *l. c.*, No. 221.

[2] Cf. above, p. 17.

[3] Who again is identical with the Gaddash of Brit. Mus. 84–2–11, 178 (Winckler, *Unters.*, p. 156, No. 6). Cf. Hilprecht, *Z. A.* VII, p. 309 *seq.*, especially note 4.

[4] Cf. Hilprecht, "Die Ergänzung der Namen zweier Kassitenkönige" in *Z. A.* VIII (in print).

[5] It is worthy of notice, that not one votive object with an inscription of a ruler of the first or second dynasty of Babylon has so far been found in Nippur. These kings concentrated their attention on the glorification of Babylon.

but by restoring the former glory of Ekur, the ancient national sanctuary in Nippur, so deeply rooted in the hearts of the Babylonian people, and by stepping forward as the champions of the sacred rights of "the father of the gods,"[1] they were able to bring about a reconciliation and a final melting together of the Cassite and Semitic elements. Supported by the influential priesthood of Nippur and dreaded as daring warriors by the discontented parties, the Cassites could mould and govern the destinies of Babylonia for nearly six hundred years, until finally they were overwhelmed by new invasions from the East and by the great national uprising in the South, which resulted in placing the native dynasty of Pashe on the throne of Babylon. The essential results to be drawn from the fifty-five votive inscriptions of the Cassite dynasty published on Plates 14–29, I have given in several articles in *Zeitschrift für Assyriologie*[2] and may therefore confine myself to the following points.

The inscriptions on Pl. 8, No. 15, and Pl. 21, No. 43, are written on the obverse and reverse of a tablet in agate. The stone tells its own story. About 2750 B. C., the patesi [3] of a city dedicated the tablet to the goddess Ninna or Ishtar " for the life of Dungi, the powerful champion, king of Ur." Afterwards, possibly about 2285 B. C., at the time of the Elamite invasion, when Kudur-Nankhundi laid hand on the temples of Akkad and carried the image of the goddess Nanâ into Elam, the tablet was also taken away and remained in the possession of the enemies until c. 1300 B. C. Kurigalzu (doubtless the second of the name [4]), after his conquest of Susa, brought it back to Babylonia and presented it to Bêltis of Nippur. For over three thousand years it lay within the walls of Ekur, until again it became the spoil of invaders of Nippur. This time it was carried far away to the modern ^{mâtu}Aharri. Perhaps a later *shar kibrat arba'im* will take it back to the resurrected sanctuary of Nippur. Kurigalzu's inscription on this tablet is of historical importance, because, for the first time, we learn from this king's own inscriptions of his successful campaign against Elam,[4] in the course of which he conquered even Susa.[5] The cuneiform text reads: 1. *Kurigalzu* 2. *shar Karuduniash* 3. *êkalla sha* ^{âlu}*Shâsha* ^{ki} 4. *sha Elamti* ^{ki} 5. *ikshud-ma* 6. *ana* ^{ilu}*Bêlit* (*NIN-LIL*) 7. *bêltishu* 8. *ana balâtishu* 9. *ikish*, " Kurigalzu, king of Karuduniash, conquered the palace of Susa in Elam and presented (this tablet) to Bêlit, his mistress, for his life."

[1] Inscription of Kadashman-Turgu, Pl. 24, No. 63, l. 1 and 2.

[2] Cf. "Bibliography," II, 9, 11, 12.

[3] This word stood apparently in one of the lost lines at the lower end of the tablet.

[4] Cf. Pinches, "An Early Tablet of the Babylonian Chronicle," in *R. P.*[2], Vol. V, p. 109, col. III, 10–18.

[5] The earliest mention of Susa in the Babylonian cuneiform literature. The absolute proof for the identity of *Shâsha* with *Shûshi* (IV R.[2] 52, 46, *b;* II R. 48, 59, *b*, and Delitzsch, *Paradies*, p. 326), *Shûshan* or *Shushun*, is impossible at present. It seems, however, scarcely possible that *êkallu sha Shâsha sha Elamti* can be anything else than שׁוּשַׁן הַבִּירָה אֲשֶׁר בְּעֵילָם (Dan. viii. 2). The name was probably pronounced *Shôsha(n)*. Cf. also p. 13, note 1 (end).

Another inscription published on the same plate, Nos. 41 and 46, was damaged
at the end of each line when the scribe cut it from the block of lapis lazuli,[1] which
Kurigalzu dedicated to Bêl.   It reads : 1. *A-na ᶦˡᵘBêl (En[-lil])*) 2. *be-el ma-ti-a-ti
be-* [*lì-²shú*] 3. *Ku-[r]i-gal-zu ri-ia-um* [*na-ram ᶦˡᵘBêlit?*][3] 4. *pa-li-ih* [*she-mu-u
ᶦˡᵘBêl?*],[3] "To Bêl, lord of the lands, his lord, Kurigalzu, the shepherd beloved
by Bêlit, he who fears (and) obeys Bêl."

The cuneiform text of the lapis lazuli disc on Pl. 23, No. 61, proves the correctness of my conjecture in *Z. A.* VII, pp. 305–318.   The fourth character of l. 3 is,
however, not as I supposed, *Ka* but *Kad*.[4]   The disc thus furnishes us the new and
interesting writing *kaddashman*[5] instead of the hitherto *kadashman*.

No. 66 and 67 of Pl. 25 are the obverse and reverse of the same fragment of an
agate ring.   The dedication on it was apparently written by one king only, who, in need
of space, inscribed both the upper and lower side of his gift.   As the remnant of the
last character of No. 66 is doubtless to be completed to *Ka-[dingir-raᵏⁱ]*, the ideogram *shar*, standing before it, must be the title of a king, whose name ended in *LIL*
(the last character of *ᵈⁱⁿᵍⁱʳEN-LIL* or Bêl).   According to our present knowledge
of the rulers of the Cassite dynasty, the name can be read either *Kudur-ᵈⁱⁿᵍⁱʳEN-
LIL*[6] (cf. No. 64) or *Kadashman-ᵈⁱⁿᵍⁱʳEN-LIL* (No. 65).   The obverse of the ring
(No. 67) contains part of a name ending in [*b*]*u-ri-ia*[*-ash*], which again can be
completed either to *Shagashalti-Buriash*, the son of *Kudur-ᵈⁱⁿᵍⁱʳEN-LIL*, or[7] to
. . . . *buriash* (No. 68, col. I, 5), the son of *Kadashman-ᵈⁱⁿᵍⁱʳEN-LIL*.   As no inscriptions of the former seem to have been found in Nippur, and the characters of
Nos. 66 and 67 resemble those of No. 68 more than of No. 64, I assign the ring to
the king mentioned in No. 68, *i. e.*, in all probability Kadashman-Buriash, who,
according to III R. 4, No. 1, was at war with an Assyrian king.[8]   The following

---

[1] Cf. Hilprecht, "Zur Lapislazuli Frage im Babylonischen," *Z. A.* VIII (in print).

[2] Brünnow, *l. c.*, 5309.   Cf. Meissner, *Beiträge zum Altbabylonischen Privatrecht*, p. 115, No. 21, 3.

[3] Uncertain ; restored according to Brit. Mus., 81, 8–30, 9, l. 8,9 (cf. Jensen, Schrader's *K. B.* III, Part 1, p. 120):
ri-'a(sic! instead of Jensen's 'u)-u na-ram ᶦˡᵘBêlit, pal-ḫu she-mu-u ᶦˡᵘ.Shamash.

[4] Brünnow, *l. c.*, 2701.   See also my "Nachtrag" in *Z. A.* VII, p. 318.

[5] This is not to be used in favor of Pinches' identification of *kaddash* with *gaddash* and *gan(kan)-dash*.   I adhere
to what I remarked in *Z. A.* VII, p. 309, note 4, until *Gaddash* or *Gandash*, the founder of the Cassite dynasty, has
actually been found written with the character *Ka* (or *Ḳa*), or the word *kad(d)ash* in Cassite proper names like Kad-
(d)ashman-Turgu, with the value *ga* (or *ḳa*).   Cf. Pl. 25, No. 68, col. I, 14, 15, *dumu sag Kad-ash-ma-an-ᵈⁱⁿᵍⁱʳEN-LIL*,
"(. . . . riash) the first son of Kadashman-EN-LIL."   My writing *dumu Ka-dá-ash-ma-an-ᵈⁱⁿᵍⁱʳ Bêl* (*Z. A.* VII, p.
309, note 3) is to be corrected accordingly.

[6] Generally read *Kudur-Bêl*.   It would be more appropriate to transliterate him Kudur-Turgu (see below).   That
he was king will be shown in my article, "Die Ergänzung der Namen zweier Kassitenkönige," *Z. A.* VIII (in print).

[7] For various obvious reasons other possibilities have been excluded as improbable.

[8] The conjecture of Delitzsch (*Kossäer*, pp. 10 *seq.;* Hommel, *Gesch.*, p. 437 *seq.*), that the Assyrian king was
Shalmaneser I, is proved by the new chronology which I am able to establish for a number of Cassite kings.   Cf.
below p. 37.

is an attempt to restore the legend according to the usual phraseology of this class of inscriptions : Obverse, [^{Dingir}En-lil lugal-a-ni-(ir) Ka-da-ash-ma-an-B]u-ri-ia-[ash], Reverse, [dumu (sag) Ka-da-ash-ma-an-^{ilu}En]-lil lugal Ka[-dingir-ra^{ki} a-mu-na-shub], "To Bêl, his lord, Kadashman-Buriash, (first) son of Kadashman-EN-LIL, king of Babylon, presented it."

The question remains to be settled, whether the name of the father of Kadash-man-Buriash is to be read Kadashman-Bêl, as has generally been done,[1] or Kadash-man-Enlil[2] or still in another way. The second reading needs no refutation. It is in itself impossible. The first seems to me at present improbable. For while there are Babylonian proper names which are composed of Babylonian words and the name of a foreign god,[3] there is no evidence that there were in use any which contain a Cassite word and at the same time the name of a Babylonian deity. The example quoted by Delitzsch[4] should be read Nazi-Shiḫu.[5] For this very reason I regard the correct pronunciation of Kadashman-^{dingir}EN-LIL as being either Kadashman-Kharbe[6] or Kadashman-Turgu,[6] in other words the Cassite king Kadashman-^{dingir}EN-LIL may represent either of the two persons. Which of the two is the more probable? There are two Cassites of the name Kadashman-Kharbe to be considered. The one was the father of Kurigalzu I.[7] As, however, there is no proof that he was a king,[8] we leave him here out of consideration, the more readily, as other reasons make his identification with Kadashman-^{dingir}EN-LIL well-nigh impossible. The other Kadashman-Kharbe is entirely out of the question,[9] as none of the six kings following the latter successively, according to List b, ends in . . . .

[1] e. g., Delitzsch, Kossäer, p. 20 ; Pinches, The Academy, September 5, 1891, p. 199, b, and last Hilprecht, Z. A. VII, p. 316.

[2] Hommel, Gesch., p. 438 : Kara-Inlil.

[3] e. g., Shuḫamuna-áḫ iddina (Delitzsch, Kossäer, pp. 18, 21, 28), Kashshù-nàdin-ahu (ib.).

[4] Kossäer, p. 18, note 1.

[5] For Cass. Shiḫu = Babyl. Marduk cf. Delitzsch, Kossäer, pp. 20, 21, 39. From the few published documents in which Nazi-Shiḫu or members of his family (cf. the passages on p. 42) are mentioned, it is evident that this Cassite family lived in Northern Babylonia and was very prominent and influential. Even Nebuchadrezzar I, shâlilu Kash-shi, treated its chief with distinction (Freibrief, col. II, 12 : Kalu Akkad). In view of the true character (Hil-precht, Z. A. p. 311, note 3) of the so-called "Cassito-Semitic vocabulary" (Delitzsch, Kossäer, p. 24 seq.), and of what has been said about the formation of proper names above, I believe Nazi-Shiḫu in V R. 44, 43a, to be the same person as the high dignitary who appears as the first witness in the "Freibrief" of Nebuchadrezzar I.

[6] For Kharbe = Bêl cf. Delitzsch, Kossäer, p. 23 ; for Turgu = Bêl cf. Hilprecht, Z. A. VII, p. 316, note 3, and the following lines above.

[7] Cf. Winckler in Z. A. II, pp. 307-311

[8] Against Delitzsch, Gesch. ("Ubersicht"), who does not hesitate to number him among the Cassite rulers.

[9] The principle stated by Winckler in Z. A. II, p. 310, l. 7-10, is correct, but his identification of Kadashman-Bêl with Kadashman-Kharbe is impossible.

riash, as is required.[1] That *Turgu* is another Cassite equivalent for the Babylonian Bêl (of Nippur), I have endeavored to show in *Z. A.* VII, p. 316, note 3. But there are other reasons for identifying Kadashman-Turgu with Kadashman-*dingir*EN-LIL: (1) The cuneiform characters of the inscriptions of Kadashman-Turgu on Plates 23, 24, are strikingly similar to those of Kadashman-*dingir*EN-LIL and especially his son (Pl. 25). (2) The son of Kadashman-*dingir*EN-LIL bears precisely the same title (Pl. 25, No. 68, col. I, 6), as Kadashman-Turgu (Pl. 24, l. 8).[2]

On Pl. 28 we meet with the first personal inscription of *Rammân-shum-uṣur*, contemporary of the Assyrian king, Bêl-kudur-uṣur. The brick legend is written in Sumerian and reads: 1. *Dingir En-lil* 2. *lugal kur-kur-ra* 3. *lugal-a-ni-ir* 4. *Dingir Rammân-shum-uṣur* 5. *siba she-ga-bi* 6. *ú-a En-lil*^{ki}*-a* 7. *sag-ush E-kur-ra* 8. *E-kur e ki-ag-gà-a-ni* 9. *shegn al-ur-ra-ta* 10. *mu-un-na-ru*, "To Bêl, lord of lands, his lord, Rammân shum-uṣur, his favorite shepherd, adorner of Nippur, chief of Ekur, built Ekur, his beloved house, with bricks."

Winckler, following Sayce,[3] latterly inclines to regard the Babylonian king "Rammân-shum-naṣir," in III R. 4, No. 5, as identical with the ruler whose inscription has just been translated.[4] This, however, is utterly impossible. Sayce and Winckler misread the name of the king mentioned in III R. According to the law underlying the formation of Babylono-Assyrian personal proper names, the cuneiform group *Rammân-MU-SHESH-IR* can only be read *Rammân-mushêshir*, "Rammân is directing (ruling)."[5] This king lived before Burnaburiash and has not even the name in common with the above-given Rammân-shum-uṣur.

---

[1] For Kadashman-*dingir*EN-LIL, himself king (Pl. 25, No. 65), was the father of another king (Pl. 25, No. 68, col. I, 16), ending in . . . . *riash* (ibid., l. 5).

[2] Besides the personal votive inscriptions of King Kadashman-Turgu, many tablets dated in his reign were found in Nippur. It is certain that he was one of the best known princes of the Cassite dynasty and ruled more than fifteen years. It seems, therefore, strange that his name, being entirely Cassite, should have been omitted by the compiler of K. 4426 (V R. 44, 21–44, *a, b*). As soon as we read the name in V R. 44, 29, *a*, Kadashman-Turgu, as I proposed above, the difficulty is removed. And, indeed, this reading finds new confirmation. All the names placed together by the compiler in V R. 44, 23–44, are purely Cassite. Therefore we are obliged to regard the ideogram in the name of Kadashman-*dingir*EN-LIL, which is explained by its Assyrian equivalent *Tukulti-Bêl* in the right column, as Cassite in the left column. That *dingir*EN-LIL was not pronounced Kharbe seems, apart from the above-given reasons, to be indicated by the fact that Kharbe in V R. 44, 33 *a* (*i.e.*, in the left column) is written phonetically *Khar-be*. From names like *Kharbi-Shiḫu* (IV R.² 34, No. 2, l. 5, 14), "Bêl (= the lord) is Marduk," we may infer that the real meaning of Kharbi was something like "lord." The use of Kharbi for the name of a certain god, resembles, therefore, closely that of *dingir*EN in the later Babylonian time (cf. Tiele, *Gesch.*, p. 538). Turgu on the other hand seems to have been *the* Bêl of the Cassites, *i. e.*, exactly corresponding in his rank to the *dingir*EN-LIL or Bêl of Nippur, the highest god of their Pantheon.

[3] *R. P.*², Vol. II, p. 207, note 1 (cf. Vol. I, p. 16).

[4] *Gesch.*, p. 102 (cf., however, pp. 88, note, and 157).

[5] Cf. *ú-shesh-she-ru*, Sanh. Kuy. 2, 31.

The brick legend on Pl. 29 was already published by Pinches in *Hebraica*, Vol. VI, pp. 55–58. I need make no apology for republishing it here, as Mr. Pinches' edition, I am sorry to say, is of little use, the cuneiform text and translation offered by him being unfortunately incorrect in all essential points. The legend was stamped " by means of a wooden block, on the brick." The stamp, however, having been carved very shallowly, the inscription, " though impressed evenly," is not very distinct on any of the many hundreds of bricks which were found.[1] Besides, the surface is covered " with a thin deposit, which adds to the difficulty of deciphering the inscription." Notwithstanding all this, I did not deem it necessary to mark any of its cuneiform characters as doubtful. My copy was made after a long and careful study of each character, and especial attention was paid to every detail. Certain cuneiform characters could not be recognized distinctly on the original except in the light immediately preceding sunrise, the best time for copying difficult cuneiform inscriptions. On the following points I am obliged to differ from Mr. Pinches:

1. Pinches: "The date of this inscription is uncertain. Judging from the style of the characters, it should be about 1500 B. C., but it may be as early as 2500 B. C." In the present writer's opinion the inscription belongs to one of the last rulers of the Cassite dynasty. For paleographic reasons it cannot be older than 1250 B. C., and in fact belongs to a king who ruled c. 1165 B. C.

2. Pinches transliterates the name of the ruler (l. 4) " Nin-Dubba," regards its bearer to be a lady, and adds, the inscription " is the only text of a queen of Mesopotamia known." Mr. Pinches should have been the more careful in introducing this regent as a female to Assyriologists. I read l. 4 *Mili-Shikhu* (see below) and regard this person as being the well-known Cassite king who ruled c. 1171–1157 B. C.

3. The first character in l. 5 is, according to Mr. Pinches, *nin*, " lady," while in reality the text gives *siba*, " shepherd."

4. Mr. Pinches reads (l. 6) *lugal Ega*, " queen of Ega," and adds, " Ega is probably another name for this city [Nippur], or for a part of it." The phrase thus misunderstood by Mr. Pinches is the very common title *lugal lig* (?)[2] *-ga*, " the powerful king."

The inscription in question reads as follows: 1. ^{Dingir}*En-lil-la(l)* 2. *lugal kur-kur-ra* 3. *lugal-a-ni-ir* 4. ^{Dingir}*Mili-*^{dingir}*Shihu* 5. *siba she-ga-bi* 6. *lugal lig* (?) *-ga* 7. *lugal ub-da tab-tab-ba* 8. *E-kur* 9. *e-ki-ag-gà-a-ni* 10. ^{shega}*al-ur-ra-ta* 11. *mu-un-na-ru*, " To Bêl, lord of lands, his lord, Mili-Shikhu, his favorite shepherd, powerful king, king of the four quarters of the earth, built Ekur, his beloved house, with bricks."

---

[1] Cf. " Table of Contents."

[2] Jensen in *Z. A.* I, p. 396, note 4.

My reasons for identifying the name in l. 4 with that of Mili-Shikhu are as follows: (1) The king must have lived after Rammân-shum-uṣur, because a few bricks of the latter[1] were found in the platform of the temple erected by him.[2] (2) Paleographic reasons point to the end of the Cassite dynasty as the date of his inscription. Apart from a certain difference of appearance between Rammân-shum-uṣur's legend and that of the king in question, the one having been inscribed, the other stamped, there is a decided similarity between the characters of the two inscriptions. (3) One of the titles (l. 5), the phraseology of the beginning (l. 1–3), and—what is especially characteristic—that of the end of the two inscriptions (l. 8–11, otherw. 10), in other words, 8 (otherw. 7) lines are absolutely identical. Hence it follows that the king in question must have ruled not long after Rammân-shum-uṣur; was possibly his successor. (4) This result is corroborated by an analysis of the first half of l. 4. The determinative dingir is not unfrequently found before the names of Cassite kings.[3] The second and third characters are to be read SHA (libbu)[4] + ba. The absence of the two inner wedges in SHA is due to the shallowness with which the characters of the stamp were carved. They are found on another (badly preserved) brick, of the same king, the legend of which was written with the hand, and differs slightly in other respects.[5] As the inscription is written in Sumerian, the syllable ba indicates that the Sumerian value of the preceding sign ended in b, in other words, was the dialectic form of a word ending in g—probably shag. As the personal proper names occurring in the later Sumerian inscriptions are, as a rule, not to be read Sumerian, but as they were actually pronounced,[6] we read the ideogram (shaba) with one of its common Semitic equivalents, kirbu, libbu, milu, etc.[7]

Only one of the Semitic ideographic values of this character fulfills the requirement of forming the beginning of one of the well-known names of the last four Cassite kings, i. e., milu or mili. As, on the other hand, there is only one Cassite king of that period who begins with Mili, I confidently believe the last group of cuneiform characters in l. 4 to be an ideogram for the god Marduk, or his Cassite equivalent Shikhu, and read the whole name accordingly Mili-Shikhu.

The following list is an attempt at restoring part of the broken List b, and giving the chronology and succession of the last twenty-four kings of the Cassite

---

[1] Together with a few of Ur-Ninib, Kurigalzu, and one of Bur-Sin I.
[2] Cf. above, p. 27, and "Table of Contents," Pl. 29, No. 82.
[3] Cf. Hilprecht in Z. A. VII, pp. 308–310.
[4] Cf. Brünnow, l. c., 7983.
[5] Cf. Vol. I, Part 2.
[6] Cf. also Jensen in Schrader's K. B. III, Part 1, p. 117, notes 6–9.
[7] Cf. Brünnow, l. c., 7985–7992.

dynasty, which ruled over Babylonia for 576 years.[1] My reasons for changing the generally accepted order of several of these kings will be found in a special article. If the date which I assigned to the first rulers of the Pashe dynasty be accepted, my chronology from Kurigalzu II to Bêl-shum-iddina II must be regarded as absolutely certain. As the rulers between Barnaburiash and Kurigalzu II are well known, it is also settled beyond doubt that Shagashalti-Buriash lived before Kurigalzu I. Nabuna'id's statements concerning the chronology of Sargon I, Hammurabi, Burna-Buriash, and Shagashalti-Buriash must be regarded as only approximate dates. The events recorded may have occurred at any time in the century before or after the year given.[2] Sennacherib's statement concerning Tukulti-Ninib's cylinder (600 years) is likewise to be understood in a broad sense.

13. Rammân-mushêshir[3] . . . . c. 1442–1423 (about twenty years).
14. Kallima(?)-Sin . . . . c. 1422–1408 (about fifteen years).
15. Kudur-Turgu[4] . . . . c. 1407–1393 (about fifteen years?).
16. Shagashalti-Buriash (his son) . c. 1392–1373 (about twenty years).
17. Kurigalzu I (son of Kadash-man-Kharbe) . . . . c. 1372–1348 (about twenty-five years).
18. Kara-indash (his older son?)[5] . c. 1347–1343 (about five years?).
19. Burna-Buriash (son of 17) . . c. 1342–1318 (about twenty-five years).
20. Kara-Khardash (son of 18) . . c. 1317–1308 (about ten years).
21. Nazi-bugash (usurper)[6]. . . . . c. 1307 (about one year).
22. Kurigalzu II (son of 19) . . . . 1306–1284 (nearly twenty-three years).
23. Nazi-Maruttash (his son) . . . 1284–1258 (twenty-six years).
24. Kadashman-Turgu (his son)[7]. . 1257–1241 (seventeen years).
25. Kadashman-Buriash (his son) . 1240–1239 (two years).
26. Is-am-me . . . . ti . . . . 1238–1233 (six years).
27. Shagashalti-Shuriash[8] . . . . . . 1232–1220 (thirteen years).

[1] I regard Peiser's doubts as to the correctness of the 576 years (Z. A. VI, p. 267 seq.) as unnecessary. Through the excavations at Nippur we are enabled to substantiate part of the statements given as to this dynasty in the list. This fact teaches us *Festina lente!*

[2] And in a sentence like "who built 700 years before Burnaburiash," we have to make even a greater allowance, as we do not know which approximate date Nabuna'id had in mind in connection with the reign of Burnaburiash.

[3] He may have lived at an earlier date.

[4] Generally read Kudur-Bêl. Cf. above, p. 32 seq.

[5] The same as Kar-indash, son-in-law of Ashur-uballit, king of Assyria. Cf. R. P.², Vol. V, p. 107, l. 5, 6, 12.

[6] Called Su-zigash in R. P.², Vol. V, p. 107, l. 10, 13.

[7] Cf. Hilprecht in Z. A. VII, p. 317 (cf. Pl. 23, No. 61). The date there assigned to Kadashman-Turgu (c. 1340 B. C.) is to be corrected according to that given above. For his identification with Kadashman-*dingirEN-LIL* see above, p. 33 seq.

[8] Cf. above, p. 11.

28. Bibe[iashu] (his son)[1]	1219–1211	(nine years).
29. Bêl-shum-iddina I	1210–1209	(one year and a half).
30. Kadashman-Kharbe	1209–1208	(one year and a half).
31. Rammân-shum-iddina	1207–1202	(six years).
32. Rammân-shum-uṣur	1201–1172	(thirty years).
33. Mili-Shikhu (his son)[2]	1171–1157	(fifteen years).
34. Marduk-abal-iddina (his son) .	1156–1144	(thirteen years).
35. Zamama-shum-iddina	1143	(one year).
36. Bêl-shum-iddina II[3]	1142–1140	(three years).

The last 24 kings = c. 303 years; the first 4 kings = 68 years; the remaining 8 kings = 205 years and 9 months (each 25–26 years in average[4]). Total, 36 kings = 576 years and nine months.

THE DYNASTY OF PASHE.[5]

The cuneiform tablet published on Pl. 30 and 31 forms a part of the collection *J. S.*, purchased by the Expedition from Joseph Shemtob[6] for the University of Pennsylvania, July 21, 1888. Unfortunately it is impossible to ascertain with certainty where the stone tablet was found.[7] In regard to its size and mineralogical character it closely resembles the "black stone of Za'aleh," to be found in I R. 66, with which it also has much in common as to its contents. Both belong to the class of the so-called *kudurru* inscriptions.[8] A piece of ground situated in the land of Kaldi, in the province of Bît-Sinmâgir (I, 1, 2), which for many years (I, 3–8) had been in possession of the family of a certain Nabû-shum-iddina (I, 15) but had been unlawfully reduced in size by Ekarra-ikîsha, at that time governor of Bît-Sinmâgir (I, 9–15), was upon the complaint of the owner (I, 16–II, 5) restored to its original extent by

[1] Identical with S. 2106, l. 9. See above, p. 11.

[2] Cf. Belser in *B. A.* II, p. 197, l. 31.

[3] Cf. *R. P.*[2], Vol. V, p. 111, l. 14; p. 112, l. 16. Cf. also below, p. 41.

[4] Such long reigns appear in no way improbable when compared with the longer reigns of fifteen rulers of the first and second dynasties of Babylon.

[5] Sayce (*R. P.*[2], Vol. I, p. 17, note 3) regards this city as identical with Isin and Patesi. Cf. II R. 53, 13a.

[6] Cf. Harper, *Hebraica* V, pp. 74–76.

[7] Cf. "Table of Contents," Pl. 30, 31.

[8] I reckon as such not only "those Babylonian documents which are inscribed on blocks of stone not always quite regularly hewn" (Belser, *B. A.* II, p. 111), but also those which, like ours and the Za'aleh stone, were kept within doors and possibly as duplicates of the "stèles," which were naturally exposed to destructive influences, so that in disputes concerning boundaries they might furnish the basis for a legal decision.

Bêl-nâdin-aplu, king of Babylon, in the fourth year of his reign (II, 6–10). The document closes with a blessing for the official who in time to come shall respect the decision (II, 11–20), and with a curse against him who shall remove the boundary again (II, 21–24).

Apart from the fact that the stone furnishes us with the name of one of the early kings of the "Sea-land," with that of a hitherto unknown province or county of the land of Kaldi,[1] and with other details of interest, it is of the greatest importance for its chronological bearings. For the following reasons, the stone must be assigned to the Pashe dynasty: (1) The cuneiform characters are those which are characteristic of the documents of that period, and especially they resemble those of the charter (*Freibrief*) of Nebuchadrezzar I.[2] (2) Ekarra-ikîsha, son of Ea-iddina, is mentioned as an official[3] both on our stone (I, 10, 11; II, 6) and on that of Za'aleh (II, 6). From this it follows that our stone belongs to about the same time as the other which bears the date of the first year of King Marduknâdinahê. (3) But we are able to fix the date of our stone even more exactly from the statement in col. I, 7–15, according to which the piece of land in question was in possession of the family of Nabû-shum-iddina until the time of Nebuchadrezzar I, but in the fourth year of King Bêlnâdinaplu was unlawfully encroached upon by the governor, Ekarra-ikîsha. The result naturally is that the stone dates from the reign of Bêlnâdinaplu, and that the latter was the immediate successor of Nebuchadrezzar I. This proves, at the same time, that the supposition made by Winckler[4] and Delitzsch,[5] that Marduknâdinahê was the immediate successor of Nebuchadrezzar I, is wrong, and that the order is rather Nebuchadrezzar I, Bêlnâdinaplu, Marduknâdinahê.

The question arises, What place must be assigned to this group of three kings in the dynasty of Pashe? This, in my opinion, can be answered with entire certainty. For although the Babylonian list[6] has been broken off at the very place where the names of the rulers of this dynasty once stood, yet the characters which remain of the last three kings serve us in solving the question. Of the five known kings of this dynasty, 1. Nebuchadrezzar I, 2. Bêlnâdinaplu, 3. Marduknâdinahê, 4. Mardukshâpikzîrim (*sic!*) (not Marduktâbikzîrim)[7] 5. Rammânapluiddina, none of them fit into the

---

[1] Delitzsch, *Paradies*, p. 202 *seq.;* Winckler, *Unters.*, p. 51 *seq.*

[2] Cf. Hilprecht, *Freibrief Nebukadnezar's I*, and V R. 55–57.

[3] On our stone he appears as "governor of Bît-Sinmâgir;" on that of Za'aleh as "governor of the city of Ishin;" so that he probably had been transferred on the accession of Marduk-nâdin-ahê, or possibly a little earlier. The previous "governor of Ishin" was Shamash-nâdin-shumu, son of Atta-ilûma (cf. *Freibrief Nebukadnezar's I*, col. ii, 17).

[4] *Gesch.*, p. 96. [5] *Gesch.*, p. 93.

[6] Winckler, *Unters.*, p. 146 *seq.*

[7] A cylinder fragment of this king, in possession of Mr. Talcott Williams, of Philadelphia, was transliterated and translated in *Z. A.* IV, 301–323. Paleographic reasons are decisive in fixing the date of this cylinder. Mr. Williams has given me his kind permission to publish the cuneiform text in the second part of the present volume. Cf. below, p. 44.

remaining characters of the last three names of the dynasty. It follows, therefore, that all the five must have reigned before these. As the kings which have been numbered 4 and 5 are known to have been successors of Marduknâdinahê, it likewise follows that Nebuchadrezzar I cannot have stood lower than the fourth place in the list. It may be safely asserted, however, that he stood in the first place, and was, therefore, the founder of the Pashe dynasty. To this two objections may be offered: (1) That the traces of the cuneiform characters which follow the number of the years in the List b do not favor the reading of *Nabû*; (2) that Sayce,[1] on the evidence furnished by the "Early Tablet of the Babylonian Chronicle,"[2] col. IV, 17, claims that place in the list for a king *Rammânu-sharra* [or *shum*][3] *-iddina*. In reply to this the following is to be said:

1. Scholars have adhered too closely to the view that the mutilated beginning of the first line of the List b contains after *ilu* traces of the sign SHU,[4] the ideogram for the god Marduk. Winckler, in his edition of the list, cuts loose from this assumption, and gives as certain only *ilu*. This variation from the carefully guarded tradition is supported by Bezold's remark[5] that "at this point the tablet is in a most lamentable condition." The latter, however, seems to recognize traces of two other wedges immediately following. But the chief problem is whether beneath the two horizontal wedges of *ilu*, there can be seen a small horizontal wedge so that the sign can be completed to the combination of *ilu* and AG,[6] the ideogram for *Nabû*. From the fact that all those who have examined the list personally are silent on this point I infer that the tablet at this place is too indistinct to permit any definite conclusion. Then, however, there is nothing in the remaining traces that forbids the reading of *Nabû* instead of Marduk.

2. From what we know from the scanty cuneiform accounts,[7] it is clear that the last years of the Cassite dynasty were a time of war and political disturbance, and that it was the weakness of its last representative which furnished the opportunity for its own overthrow and for the rise of the house of Pashe. No matter what verb may have stood in the effaced passage *R. P.*[2], Vol. V, p. 112, l. 16,[8] the supposition

---

[1] *R. P.*[2], Vol. V, p. 112, note 1.

[2] *R. P.*[2], Vol. V, pp. 106–114.

[3] The reading of the middle character seems to be doubtful. Mr. Pinches would render a great service to Assyriologists by publishing the exact cuneiform text at an early date.

[4] Brünnow, *l. c.*, 10834.

[5] *Z. A.* IV, p. 317, note 1.

[6] Brünnow, *l. c.*, 2786. Cf. Hommel, *Gesch.*, p. 448.

[7] Cf. especially *R. P.*[2], Vol. V, pp. 111, 112, l. 14–22.

[8] I favor *umashshir*, "he left," instead of "he renounced" or "abdicated" (Pinches). Cf. however, Tiele, *l. c.*, p. 165.

of Sayce, that line 17 contains the name of the second king of the Pashe dynasty, seems to me improbable, since the same Elamite king, *Kidin-Khutrutash*,[1] who already had attacked Akkad in the time of Bêlshumiddina, is again the assailant in this passage. If Sayce were right, this Elamite would have made his second incursion into Akkad about twenty years after the first. This in itself is possible, but it is made less probable by the expression " Rammânu-shum-iddina returned," which apparently connects this section closely to that which precedes. Besides it will be noticed that Rammânu-shum-iddina does not bear the title of king, as Bêlshumiddina. It seems more probable, therefore, to see in Rammânu-shum-iddina, the unfortunate son (or possibly another relative) of Bêlshumiddina, who "returned" from the place to which Bêlshumiddina or his family had fled, in order to take possession of the throne as his lawful inheritance.

This leads me to the discussion of the reasons for regarding Nebuchadrezzar I as the founder of the Pashe dynasty.

1. It needs no proof that at a time when a country is harried by a powerful enemy,[2] and a descendant of illustrious ancestors puts forward claims to the crown, which are based on historic rights, a usurper who is to found a new dynasty must distinguish himself by eminent courage and ability. Such an able ruler, who, according to our present knowledge, surpassed in preëminence all the other kings of his dynasty, Nebuchadrezzar I is certified to have been. He conducted successfully the wars against Elam, the hereditary enemy of Babylon in the East, turned his arms victoriously against the North by "casting down the mighty Lulubæan," and marched, as no other Babylonian king for centuries had ventured, conquering into Syria.

2. It is worthy of notice that both the documents bearing his name are written in connection with his successful conflict with Elam. His wars with this country, therefore, must have been especially important, perilous and of long duration.[3] Since we have learned from Pinches' recent publication of the Babylonian Chronicle (col. IV, l. 14–22) that the Elamites took advantage of the weakness of the last Cassite king to devastate Northern and Southern Babylonia, the campaigns of Nebuchadrezzar I against Elam become of especial significance. As a usurper he manifestly was able to hold his position only by rendering the Elamites harmless and by defeating them on their own soil, thus " avenging Akkad,"[4] and restoring quiet and peace to his own country.

[1] This and not *Khutru ana* or *Khutrudish* (Pinches, *l. c.*, pp. 111–113) is the probable reading. For the value *tash* of the character in question see Hilprecht in *Z. A.* VII, pp. 309, 310, 314. The name means "subject (servant) of the god Khutrutash " (cf. god Marútash).

[2] R. P.², Vol. V, pp. 111 *seq.*

[3] Winckler, *Gesch.*, p. 96.

[4] Hilprecht, *Freibrief*, col. I, 13.

A. P. S.—VOL. XVIII. F.

3. Nebuchadrezzar I bears titles which differ entirely from those at that time characteristic of the rulers of Babylonia. He calls himself, in the manner of the Egyptians, *Shamash mâtishu*, "the Sun of his land;" or *mushammiḫu nishishu*, "he who makes prosperous his people;" *nâṣir kudûrêti, mukînu ablê*,[1] "he who protects the boundaries, establishes (measured) tracts of land;" *shar kinâti sha dîn mîshari idinu*, "the king of the right, he who judges a righteous judgment;" all are titles which probably refer to the fact that just before the reign of Nebuchadrezzar I there was in Babylonia a time of profound misery, when the land did not enjoy sunshine, and when the peaceful possession of well-defined property was impossible, as the violence of the stronger superseded law and order, while, at the same time, the boundaries of the empire were constantly invaded by powerful enemies; in other words, anarchy as we know it existed in Babylonia at the close of the reign of Bêlshum-iddina. The significant title, *shâlilu Kashshi*, "the conqueror of the Cassites," acquires doubtless, in this connection, the significance of an allusion to the circumstance that it was he who had achieved the restoration of the Semitic element through the overthrow of the Cassite dynasty.[2]

4. The boundary stone IV R.[2], 38, which is dated in the time of Merodachbaladan I, mentions the house (I, 10) and the son (II, 34, 35) of a certain Nazi-Shikhu, while in the "Freibrief" of Nebuchadrezzar I, a certain Nazi-Shikhu is named as a high dignitary, *kalu Akkad*. In view of the rare occurrence of this name in Babylonian literature [3] it is natural to regard the two bearers of the same name as identical. This identification, however, is possible only if Nebuchadrezzar I reigned not long after Merodachbaladan I,[4] i. e., if he, as founder of the Pashe dynasty, came into power some four years after the latter's death.

---

[1] I formerly transliterated this word *aplê* (as Peiser still does in Schrader's *K. B.* III, Part 1, p. 164). But since 1886 I have changed my view and substituted the above. As the word stands parallel. to *kudûrêti*, it must have a similar meaning. In spite of *naḫbalu*, II R. 22, 29, *b. c.*, *ablê* is to be compared with the Hebrew, חֶבְלִי which, in view of the Ethiopic and Arabic *ḥabl* has ḥ. Cf. also Delitzsch, *Wörterbuch*, p. 37, no. 30. In view of the title above quoted it does not seem improbable that Nebuchadrezzar I assumed his highly significant name, "Nebo, protect the boundary," only after his usurpation. Another interpretation of the name, "Nebo, protect (thy) servant," has recently been offered by Jäger (*B. A.* I, 471, note *). But where is the "thy"? The proper names *kudurru* and *kidinnu*, quoted by Jäger, (*l.c.*), are not to be regarded as exclamations but as abbreviations of originally longer names. As the middle part of the name of Nebuchadrezzar is written either *kudurru* or *kudurri* (Bezold, *Babylonisch-Assyrische Literatur*, p. 126), or *kudurra* (Pl. 32, col. II, 7, of the present volume), it cannot mean "my boundary," as I formerly interpreted (*Freibrief*, p. viii, note 1), but "the boundary." Cf. my remarks in *The Sunday School Times*, February 20, 1892, p. 115, note 3.

[2] Cf. Hommel, *Gesch.*, p. 451.

[3] Cf. col. VI, 18 of the boundary stone (published by Belser in *B. A.* II, pp. 171–185), which furnishes us data from the time of the kings *Ninib-kudûri-uṣur* and *Nabû-mukin-aplu*. For my transliteration and the formation of the name, cf. above, p. 33 and note 5.

[4] For as the son of Nazi-Shikhu who appears as a witness under Merodachbaladan I, was already in possession of the important office of a *sukallu*, his father must have been advanced in years.

5. The second king of the Pashe dynasty, according to List b, reigned only six years. And indeed, while the titles and conquests of Nebuchadrezzar I in his "Freibrief" imply a comparatively long reign, there are indications that his immediate successor, Bêlnâdinaplu, ruled but a short time. This does not necessarily follow from the circumstance that the document on Plates 30 and 31 is dated in the fourth year of his reign; but from the fact that Ṭâb-ashâp-Marduk,[1] son of Esagilzêr,[2] already mentioned under Nebuchadrezzar I as governor of Ḫalwân, appears again as sukallu in the first year of Marduk-nâdin-aḫê, i. e., about twenty years later; for it is very unlikely that the same person occupied a high and responsible position under three successive kings, if both of the former two had reigned a long period.

6. Finally this assumption enables us in the simplest way to dispose of certain chronological difficulties, upon which I cannot enter into details here (cf. e. g. Z. A. III, p. 269).

The statement of Sennacherib[3] furnishes us with a definite datum for the chronology of the Pashe dynasty. As it seems most natural to connect the carrying off of the images of the gods of Ekallâti, with Marduknâdinaḫê's victory over Assyria, in the tenth year of his reign,[4] we obtain 1107 B. C. as the tenth year of that king's rule, and 1116 B. C. as the year of his accession to the throne. In accordance with what has been said above, Nebuchadrezzar I reigned 1139–1123 B. C.,[5] and Bêl-nâdin-aplu in 1122–1117 B. C.

A word remains to be said as to the length of the period covered by the Pashe dynasty. That the reading of seventy-two years which have been generally assigned to it is impossible, Peiser has shown beyond question by a very simple calculation.[6] The number of twelve years for the seventh king of this dynasty, assumed by Tiele

---

[1] The reading *Ṭabu-ri'êu-Maruduk*, "A beneficent king is Marduk," preferred by Tiele (*Gesch.*, p. 161, note 1), instead of that given above (and first proposed by Oppert and Ménaut in *Documents Juridiques*), needs no refutation. *Ṭâb-ashâp-Marduk* is the only possible one and means "Good is the exorcism of Marduk." The *Caillou de Michaux* upon which *Ṭâb-ashâp-Marduk*, apparently not so far advanced in years, likewise appears, belongs to the reign of Nebuchadrezzar I or of Bêlnâdinaplu (cf. Tiele, *l. c.*, p. 161, and Hommel, *Gesch.*, pp. 454, 459).

[2] That *Esagilzêr* is identical with the *Ina-Esagilzêr* of the Za'aleh stone (col. II, 12), was shown in my commentary on the "Freibrief Nebukadnezar's I," in 1882, which at the time was not printed because of a two years' illness. At present the proof of their identity is unnecessary. Cf. *Eulbar-shurki-iddina*, III R. 43, col. I, 29, and *Ina-Eulbar-shurki-iddina*, V R. 60, col. I, 29. Cf. also Delitzsch, *Kossäer*, p. 15 (cf. however *Gesch.*, "Übersicht"). To a different effect Jeremias in *B. A.* I, pp. 270, 280; and Peiser in Schrader's *K. B.* III, Part 1, p. 177.

[3] *Bavian*, 48–50. "Rammân and Sala, the gods of the city of Ekallâti, which Marduknâdinaḫê, king of Akkad, at the time of Tiglath-Pileser, king of Assyria, carried off and brought to Babylon, I carried out of Babylon 418 years later, and brought them back to Ekallâti, to their place," *i. e.*, in the year B. C. 689, when Sanherib conquered Babylon.

[4] Cf. III, R. 43, col. I, 5, 27, 28.

[5] This calculation confirms strikingly the year 1130 B. C., which I gave as the approximate date of his "Freibrief" in 1883.

[6] *Z. A.* VI, p. 268 *seq.*

(*l. c.*, p. 111) and favored by Delitzsch,[1] finds no support in Winckler's edition and besides does not suffice to solve the chronological difficulty. As according to Peiser (*l. c.*) the passage is much effaced,[2] and as his proposed reading, $60 + 60 + 12 = 132$ years, is the most simple and probable[3] solution of the existing difficulty, I accept it and accordingly construct the following table:

1. Nebuchadrezzar I, . . 1139–1123 (seventeen years).
2. Bêl-nâdin-aplu, ... 1122–1117) (six years).
3. Marduk-nâdin-ahê, . . 1116–c. 1102 (c. fifteen, at least ten, years).
4. Marduk-shâpik-zîrim,[4] ⎫
5. Rammân-aplu-iddina, ⎬ c. 1101–1053 (forty-nine years).
6-7. Two missing kings, ⎭
8. . . . . , . 1052–1031 (twenty-two years).
9. Marduk-bêl . . . . . , . 1030–1029 (one year and six months).
10. Marduk-zêr . . . . . , . 1029–1016 (thirteen years).
11. Nabû-shum . . . . . , . 1016–1007 (nine years).

Total one hundred and thirty-two years and six months.

[1] "Anhang" to his *Geschichte*.

[2] It is to be regretted that Winckler has not indicated the actual condition of the passage by shading the effaced portions of the characters.

[3] Cf. also Winckler, *Gesch.*, p. 329, note 17. Another possibility (that $60 + 10 + 10 + 2 = 82$ stood originally there) is less probable for various reasons.

[4] This name has been transliterated *Marduk-shapik-zêr-mâti* (Tiele, *Gesch.*, p. 155; Delitzsch, *Gesch.*, "Übersicht") or *Marduk-shapik-kul-lat* (Winckler, *Gesch.*, p. 98). I regard both transliterations as incorrect, and would substitute that given above for the following reasons: (1) The cylinder fragment published by Dr. Jastrow (cf. above, p. 31, note 7) was unfortunately misunderstood by the latter and misread in various passages. Having examined the fragment carefully, I find that the old Babylonian character transliterated *ta* by Jastrow is distinctly the sign *sha* in the form so characteristic for the documents of the Pashe dynasty. The name can only be read *Marduk-shapik-zi-ri-im*. (2) This correct reading is important in connection with the transliteration of the name of Rammân-aplu-iddina's predecessor. It is in itself improbable that two rulers of a Babylonian dynasty of eleven kings bore names almost (if not wholly) identical. The thought forces itself upon our mind that Marduk-shâpik-zîrim is the same person as the king whose name was heretofore generally read Marduk-shâpik-zêr-mâti. That at least these two names are identical is certain. The last character of the latter name (*MAT*, Brünnow, *l. c.*, 7386) was either erroneously read by the Assyriologists who copied the so-called "synchronistic history," or by the Assyrian compiler who used a Babylonian original, instead of the character *RIM* (Brünnow, *l. c.*, 8867). For it is well known among Assyriologists that the two characters are nearly identical in the later-middle and the latest periods of Babylonian cuneiform writing. In consideration of this fact, and in view of the phonetic writing *zi-ri-im* on the cylinder fragment, I unhesitatingly read the name in question either phonetically *Marduk-shapik-zir-rim*, or ideographically (plus phonetic complement) *Marduk-shapik-zîrim(-rim)*. The king, Marduk-tâbik-zîrim, introduced by Dr. Jastrow and accepted by Peiser (Schrader's *K. B.* III, Part 1, p. 162 *seq.*) as an hitherto unknown ruler of the Pashe dynasty thus disappears. As to my other corrections of certain readings offered by Dr. Jastrow in connection with the cylinder in question cf. "Sprechsaal" in one of the next numbers of *Z. A.*

BIBLIOGRAPHY OF THE EXPEDITION.

I. JOHN P. PETERS.

1. Letter on the Babylonian Expedition : *The American Journal of Archæology* VII, pp. 472–475.
2. A Brief Statement concerning the Babylonian Expedition sent out under the auspices of the University of Pennsylvania : *Proceedings of the American Oriental Society*, April 21–23, 1892, pp. CXLVI–CLIII.
3. Notes on Mürdter-Delitzsch's Geschichte : *Zeitschrift für Assyriologie* VI, pp. 333–339.
4. A Few Ancient Sites, I and II : *The Nation* 1889, May 23, p. 423, and May 30, pp. 442, 443.
5. From Niffer to Tello, I and II : *ibidem* 1889, July 25, pp. 69, 70, and August 1, pp. 90–92.
6. Zenobia, Palmyra, and the Arabs : *ibidem* 1890, April 3, pp. 276, 277.
7. A Misrepresented Ruin : *ibidem* 1891, May 7, pp. 375–377.

II. H. V. HILPRECHT.

1. Keilinschriftliche Funde in Niffer : *Zeitschrift für Assyriologie* IV, pp. 164–168.
2. Aus einem Briefe desselben an C. Bezold : *ibidem* IV, pp. 282–284.
3. Die jüngsten Ausgrabungen in Babylonien : *Kölnische Zeitung* 1889, June 30, No. 179.
4. Neue Forschungen in Babylonien : Luthardt's *Evangelisch Lutherische Kirchenzeitung* 1889, June 14, pp. 568, 569.
5. The Mouth of the Nahr-el-Kelb : *The Sunday School Times* 1889, Vol. XXXI, No. 11, p. 163.
6. Die Inschriften Nebukadnezar's im Wadi Brissa : Luthardt's *Zeitschrift für kirchliche Wissenschaft und kirchliches Leben* 1889 IX, pp. 491–498. Compare also *The Sunday School Times* 1889, Vol. XXXI, No. 35, pp. 547, 548 : The Inscriptions of Nebuchadnezzar in the Wady Brissa.
7. The Shaykh of Zeta : *The Sunday School Times* 1890, Vol. XXXII, No. 10, pp. 147, 148.
8. Babylon : *ibidem* 1892, Vol. XXXIV, No. 20, pp. 306–308.
9. Die Votivinschrift eines nicht erkannten Kassitenkönigs : *Zeitschrift für Assyriologie* VII, pp. 305–318.
10. König Ini-Sin von Ur : *ibidem* VII, pp. 343–346.
11. Die Ergänzung der Namen zweier Kassitenkönige : *ibidem*, in print.
12. Zur Lapislazuli Frage im Babylonischen : *ibidem*, in print.

III. ROBERT FRANCIS HARPER.

1. Babylonian Letter.—The Joseph Shemtob Collection of Babylonian Antiquities, recently purchased for the University of Pennsylvania : *Hebraica* V, pp. 74–76.
2. The Kh. Collection of Babylonian Antiquities belonging to the University of Pennsylvania : *ibidem* VI, pp. 59, 60.
3. The Destruction of Antiquities in the East : *ibidem* VI, pp. 225, 226.
4. Three Contract Tablets of Ashuritililani : *ibidem* VII, p. 79.
5. A Visit to Zinjirli : *The Old and New Testament Student* VIII, pp. 183, 184.
6. A Visit to Carchemish : *ibidem* IX, pp. 308, 309.
7. Down the Euphrates Valley I–III : *ibidem* X, pp. 55–57 ; 118, 119 ; 367, 368.
8. The Expedition of the Babylonian Exploration Fund, A. B. C. : *ibidem* XIV, pp. 160–165 ; 213–217 ; XV, pp. 12–16 ; D. : *The Biblical World* I, pp. 57–62.
9. Aus einem Briefe desselben an C. Bezold : *Zeitschrift für Assyriologie* IV, pp. 163, 164. Compare also *Hebraica* VIII, pp. 103, 104 : A-bi-e-shu-' = Ebishum.
10. The Site of Old Baghdâd : *The Academy* 1889, February 23, p. 139.
11. A New Babylonian Contract : *ibidem* 1889, April 20, p. 274.

IV. THEOPHILUS G. PINCHES (based upon communications from Dr. Peters and Dr. Harper).

1. An Early Babylonian Inscription from Niffer : *Hebraica* VI, pp. 55–58.
2. The Discoveries of the American Expedition to Babylonia : *The Academy* 1891 September 5, p. 199. Compare also his note "Kadashman :" *ibidem* 1891, September 12, p. 221.

TABLE OF CONTENTS.

Part I, Plates 1–35 and I-XV.

ABBREVIATIONS.

c., circa ; **C. B. M.**, Catalogue of the Babylonian Museum, University of Pennsylvania ; **col.**, column(s) ; **d.**, diameter ; **Dyn.**, Dynasty ; **E.**, East ; **fragm.**, fragment(ary) ; **h.**, height ; **Inscr.**, Inscription ; **l.**, length ; **li.**, line(s) ; **m.**, meter ; **N.**, North ; **Nippur I, II, III**, etc., refers to the corresponding numbers on Plate XV ; **No.**, number ; **Nos.**, numbers ; **N. P.**, Notebook of Dr. Peters made on the ruins of Nippur during the second year's excavations ; **Obv.**, Obverse ; **orig.**, original(ly) ; **p.**, page ; **Pho.**, Photograph ; **Pl.**, Plate ; **Rev.**, Reverse ; **S.**, South ; **Sq.**, Squeeze ; **T.**, Temple of Bêl ; **th.**, thick(ness) ; **W.**, West ; **w.**, width ; **Z.**, Ziqqurratu ; **Z. A.**, Zeitschrift für Assyriologie.

Measurements are given in centimetres. Whenever the object varies in size, the largest measurement is given

I. AUTOGRAPH REPRODUCTIONS.

OLD BABYLONIAN INSCRIPTIONS

PLATE.	TEXT.	DATE.	DESCRIPTION.
5	6	Âl-usharshid.	Fragm. of a vase in reddish numulite limestone, h. 16.5, d. 18 (of hole 4.4). *Nippur* III, same place as Pl. 4, No. 5. Inscr. orig. 11.75 × 7.05, 6 li. C. B. M. 8888. The text has been restored after No. 5. Cf. Pl. IV, 13.
5	7	Âl-usharshid.	Fragm. of a white marble vase, h. 21, d. 16.4 at the base, 11.2 at the centre. *Nippur* III, same place as Pl. 4, No. 5. Inscr. 4.8 × 5.4, 3 li. C. B. M. 8870. Cf. Pl. V, 14.
5	8	Âl-usharshid.	Fragm. of a white marble vase, orig. h. 6, d. 14.5. *Nippur* III, same place as Pl. 4, No. 5. Inscr. (same as Pl. 5, No. 7) 3.2 × 3.8, 3 li. C. B. M. 8839.
5	9	Âl-usharshid.	Fragm. of a white marble vase, orig. h. 13.5, d. 15 (of hole 6.3). *Nippur* III, same place as Pl. 1, No. 1. Mark on the bottom, 2.4 × 2.6. Same inscr. as Pl. 5, No. 7. N. P.
5	10	Âl-usharshid.	Fragm. of a diorite vase, 7.35 × 2.9 × 0.8, orig. d. 22.2. *Nippur* III, same place as Pl. 4, No. 5. Inscr. 3, orig. 11 li. C. B. M. 8842.
6	11	Same Period.	White marble tablet, Obv. flat, Rev. rounded, 11.3 × 7.2 × 2.65. *Nippur*, apparently from the N. W. extremity of V in the neighborhood of Pl. 3, No. 4 (cf. Hilprecht in Z. A. IV, pp. 282–284). Inscr. 8 (Obv.) + 7 (Rev.) = 15 li. C. B. M. 8757. Copied by myself on the ruins of Nippur, April 8, 1889.
7	12	Same Period.	Fragm. of a large vase in white marble, 10 × 12.5 × 6.2. Presumably neighborhood of *Babylon*. Inscr. 2 col., 8 li. C. B. M. 1128.
7	13	c. 3000 B.C.	Fragm. of a slab in compact limestone, 12.8 × 7.35 × 5.55. *Nippur* III, inside of the great S.E. temple wall. Inscr. 3 col., 15 li. C. B. M. 8841.
8	14	Ur-Gur.	Basalt tablet, Obv. flat, Rev. rounded, lower left corner wanting, 12.25 × 5.58 × 2.2. Northern Babylonia, probably *Ursag-Kêsh*. Inscr. 8 (Obv.) + 1 (Rev.) = 9 li. C. B. M. 841.
8	15	Dungi.	Agate tablet, bored lengthwise, both sides convex, lower part wanting, 4.4 × 4.3 × 0.8. *Nippur* III, in a chamber on the edge of the canal outside of the great S.E. wall of T. Obv. Inscr. 8 li. C. B. M. 8598. For Rev. see Pl. 21, No. 43.
9	16	Dungi.	Soapstone tablet, Obv. flat, Rev. rounded, 8.6 × 5 × 1.88. Babylonia, probably *Muqayyar*. Inscr. 6 (Obv.) + 2 (Rev.) = 8 li. C. B. M. 842.
9	17	Ishme-Dagân	Fragm. of a slab in diorite, 8.1 × 10.5 × 5.6. *Nippur* III, S. of Z. Inscr. 3 col., 3 + 2 + 2 = 7 li. C. B. M. 3243.
10	18	Ur-Ninib.	Fragm. of a brick of baked clay, brown, 32 (orig.) × 23 (fragm.) × 8.4 (orig.). *Nippur* III, found out of place in a later structure on the S.E. side of Z. (cf. Pl. 29, No. 82; Pl. 13, No. 22; Pl. 20, No. 38). Inscr. (written) 23.3 × 10.65, 13 li. C. B. M. 9021. Cf. IV, R. 35², No. 5.
11	19	Bur-Sin I.	Fragm. of a brick of baked clay, brown, 30.5 (fragm.) × 20 (fragm.) × 6.5 (fragm.). *Nippur* III, found out of place, same place as Pl. 10, No. 18. Inscr. (stamped) 22.5 × 10.5, 10 li. C. B. M. 8642.
12	20	Bur-Sin II.	Door socket in diorite, an irregular cube, c. 19 each side. *Nippur* III, in a small shrine outside of the great S.E. wall of T. Inscr. 15.4 × 13.4, 2 col., 11 + 6 = 17 li. C. B. M. 8838.

PLATE.	TEXT.	DATE.	DESCRIPTION.
13	21	Bur-Sin II.	Door socket in diorite, 33 × 28 × 23. *Nippur* III, same place as Pl. 11, No. 19. Inscr. around the hole, 23.5 × 5.35, 17 li. Sq. On the bottom at the edge is the same inscr. as Pl. 14, Nos. 23-25 (cf. also Pl. 1, No. 1).
13	22	Bur-Sin II.	Brick of baked clay, light brown, very soft, covered with bitumen, 30 × 30 × 6.5. *Nippur* III, same place as Pl. 11, No. 19. Inscr. (written) 5.97 × 5.3, 2 li. Sq. The inscription is generally repeated three or four times on the same brick (edges and sides).
14	23-25	Gande.	Large unhewn blocks of white marble and reddish granite, varying in d. from 25-60. *Nippur* III, approximately same place as Pl. 1, No. 1. Inscr. 6 × 5.3 ; 7 × 6.2 ; 6.5 × 7.7 ; each 3 li. Sq.
15	26	c. 2250 B.C.	Cream-colored soapstone tablet, Rev. broken off, 4.85 × 4 × 0.8. Presumably neighborhood of *Babylon*. Inscr. 8 li. C. B. M. 103.
15	27	Hammurabi.	Fragm. of an ornamented soapstone stamp in the shape of a vase, h. 13.3, d. 12.2 at the bottom, 8.7 at the centre. Presumably neighborhood of *Babylon*. Inscr. (on the bottom) 8 li. C. B. M. 1126. Cf. Pl. IX, 20.
15	28	Cassite Dyn.	Lapis lazuli disc, d. 1.7. The thickness of this class of inscribed objects found at the same place, if not expressly stated in the following lines, varies from 0.2 to 0.8 cm. *Nippur* III, same place as Pl. 8, No. 15. C. B. M. 8685.
15	29	Cassite Dyn.	Agate cameo, d. 1.55. *Nippur* III, same place as Pl. 8, No. 15. C. B. M. 8687.
15	30	Cassite Dyn.	Lapis lazuli disc, d. 1.6. *Nippur* III, same place as Pl. 8, No. 15. C. B. M. 8721.
15	31	Cassite Dyn.	Agate cameo, bored lengthwise, 1.7 × 1.9. *Nippur* III, same place as Pl. 8, No. 15. C. B. M. 8723.
15	32	Cassite Dyn.	Lapis lazuli tablet, bored lengthwise, 1.65 × 1.8. *Nippur* I, apparently out of place, in a gully on the surface. C. B. M. 8720.
16	33	Burna-Buriash.	White marble mortar ; an uninscribed portion is broken from its side, h. 14.4, d. 12.8. Presumably neighborhood of Babylon. Inscr. 31.5 × 11.25, 27 li. C. B. M. 12. Cf. Pl. IX, 21.
17	33	Burna-Buriash.	The same, continued.
18	34	Burna-Buriash.	Ivory knob of a sceptre (or conventionalized form of a phallus), top rounded, base flat, round hole in the centre, h. 3.5, d. 5.9 at the top, 6.2 at the bottom. *Nippur* III, same place as Pl. 8, No. 15. Inscr. 5.8 × 2.42, 5 li. C. B. M. 8730. Cf. Pl. X, 23.
18	35	Kurigalzu.	Tablet in feldspar (mottled dark brown and gray), upper (inscribed) surface convex, lower flat, 3 × 12.2 × 0.9. *Nippur* III, same place as Pl. 8, No. 15. Inscr. 2 li. C. B. M. 8600.
18	36	Kurigalzu.	Irregular block of lapis lazuli, upper part inscribed, 5.1 × 9.25 × 5. *Nippur* III, same place as Pl. 8, No. 15. Inscr. 3.38 × 4.48, 6 li. C. B. M. 8599. Cf. Pl. XI, 25.
19	37	Kurigalzu.	Door socket in white marble with red veins here and there, 46.5 × 43.8 ×22. *Nippur* III, on the N.E. side of T. near the outer wall. Inscr. on both sides of the hole, 11 li. intended, but only 7 li. inscribed, 14.3 × 14.3. Copied by myself on the ruins of Nippur, April 6, 1889.

PLATE.	TEXT.	DATE.	DESCRIPTION.
20	38	Kurigalzu.	Fragm. of a brick of baked clay, brown, 32 (orig.) × 17 (fragm.)× 7 (orig.). *Nippur* III, found out of place in a later structure of the inner wall of Z. (cf. Pl. 29, No. 82; Pl. 10, No. 18). Inscr. 13.5 × 6, 9 li, stamped on the edge; the space being too small, a portion of the last character of each line is wanting. C. B. M. 8636.
20	39	Kurigalzu.	Fragm. of an axe in imitation of lapis lazuli, 9 × 6.3 × 2.7. *Nippur* III, same place as Pl. 8, No. 15. Inscr. 7 li. C. B. M. 9462. Cf. Pl. XI, 26.
21	40	Kurigalzu.	Fragm. of an axe in imitation of lapis lazuli, 5 × 6.35 × 1.5. *Nippur* III, same place as Pl. 8, No. 15. Inscr. 4 li. C. B. M. 8661.
21	41	Kurigalzu.	Fragm. of a lapis lazuli tablet, 1.7 × 1.7. *Nippur* III, same place as Pl. 8, No. 15. Inscr. 3 li. C. B. M. 8662. Originally it formed part of No. 46.
21	42	Kurigalzu.	Fragm. of a lapis lazuli tablet, 1.8 × 1.2. *Nippur* III, same place as Pl. 8, No. 15. Inscr. 2 li. C. B. M. 8663.
21	43	Kurigalzu.	Agate tablet. Rev. of Pl. 8, No. 15. Inscr. 9 li.
21	44	Kurigalzu.	Fragm. of a turquoise tablet. Obv. flat, Rev. rounded; hole bored nearly perpendicular to the lines of the Obv.; 3.4 × 3.4 × 0.8. *Nippur* III, same place as Pl. 8, No. 15. Inscr. 4 li. C. B. M. 8664.
21	45	Kurigalzu.	Lapis lazuli tablet, with two holes, 2 × 2.6. *Nippur* III, same place as Pl. 8, No. 15. Inscr. 2 li. C. B. M. 8665.
21	46	Kurigalzu.	Two fragm. of a lapis lazuli tablet, 3.65 × 7.25. *Nippur* III, same place as Pl. 8, No. 15. Inscr. 4 li. In cutting the tablet from the original block of lapis lazuli the last characters of each line were lost. C. B. M. 8666. The copy has been made from an electrotype, on which the space between the two fragments was given too small (cf. No. 41).
22	47	Kurigalzu.	Nine fragm. of a lapis lazuli tablet, 5.1 × 6 × 0.7. *Nippur* III, same place as Pl. 8, No. 15. Inscr. 6 li. C. B. M. 8667.
22	48	Kurigalzu.	Lapis lazuli tablet, hole bored near the top parallel with the lines. 2.8 × 3.45. *Nippur* III, same place as Pl. 8, No. 15. Inscr. 5 li. C. B. M. 8668.
22	49	Kurigalzu.	Lapis lazuli disc, hole bored near the centre parallel with the lines d. 2.5. *Nippur* III, same place as Pl. 8, No. 15. Inscr. 3 li. N. P.
22	50	Kurigalzu.	Fragm. of an agate ring, d. 1, w. 0.9. *Nippur* III, same place as Pl. 8, No. 15. Inscr. 5 li. C. B. M. 8669. The ring originally formed the beginning of a votive cylinder (c. 2.6 cm. long), which was afterwards cut in 3 pieces, each thus forming a ring. For the centre part see Pl. 26, No. 74. The last part has not been found.
22	51	Kurigalzu.	Agate cameo, 3.2 × 2.4. *Nippur* III, same place as Pl. 8, No. 15. Inscr. 4 li. N. P.
22	52	Kurigalzu.	Fragm. of an agate cameo, 1.7 × 1.2. *Nippur* III, same place as Pl. 8, No. 15. Inscr. 2 li. C. B. M. 8670.
22	53	Nazi-Maruttash.	Fragm. of a lapis lazuli disc, d. 2.97. *Nippur* III, same place as Pl. 8, No. 15. Inscr. 6 li. N. P.

PLATE.	TEXT.	DATE.	DESCRIPTION.
22	54	Nazi-Maruttash.	Lapis lazuli disc, d. 2.05. *Nippur* III, same place as Pl. 8, No. 15. Inscr. 5 li. N.P.
22	55	Nazi-Maruttash.	Fragm. of an axe in imitation of lapis lazuli, 4.7 × 4.6 × 1.7. *Nippur* III, same place as Pl. 8, No. 15. Inscr. 4 li. C. B. M. 8671.
23	56	Nazi-Maruttash.	Magnesite knob of a sceptre (or conventionalized form of a phallus), top rounded, base flat, round hole in the centre, h. 5.2, d. 6.9. *Nippur* III, same place as Pl. 8, No. 15. Inscr. around the top, badly effaced. C. B. M. 8728. Cf. Pl. X, 24.
23	57	Nazi-Maruttash.	Magnesite knob of a sceptre (or conventionalized form of a phallus), top slightly rounded, base flat, hole in the centre (round above, square below), h. 5.2, d. 6.1. *Nippur* III, same place as Pl. 8, No. 15. Inscr. around the top, badly effaced. C. B. M. 8727. Cf. Pl. X, 22.
23	58	Nazi-Maruttash.	Fragm. of a lapis lazuli disc, d. 4.4. *Nippur* III, same place as Pl. 8, No. 15. Inscr. 5 li. (orig. 8). N. P.
23	59	Kadashman-Turgu.	Fragm. of a lapis lazuli disc, d. 3.7. *Nippur* III, same place as Pl. 8, No. 15. Inscr. 6 li. (orig. 7). N. P.
23	60	Kadashman-Turgu.	Fragm. of a lapis lazuli disc, d. 2.55. *Nippur* III, same place as Pl. 8, No. 15. Inscr. 4 li. (orig. 5). C. B. M. 8722.
23	61	Kadashman-Turgu.	Lapis lazuli disc, d. 3.55, th. 0.35. Place unknown, probably *Nippur*. Inscr. 8 li. Original in the Museum of Harvard University, Cambridge, Mass. Cf. Lyon in *Proceedings of the American Oriental Society*, May, 1889, pp. cxxxiv–cxxxvii, and Hilprecht in Z. A. VII, pp. 305–318.
23	62	Kadashman-Turgu.	Lapis lazuli disc, d. 2.7. *Nippur* III, same place as Pl. 8, No. 15. Inscr. 5 li. C. B. M. 8673.
24	63	Kadashman-Turgu.	Irregular block of lapis lazuli, 17.5 × 11 × 9. *Nippur* III in a room in the mounds S. of T. Inscr. 16.4 × 9.5, 20 li. Sq.
25	64	Kudur-EN-LIL.	Lapis lazuli disc, d. 2.5. *Nippur* III, same place as Pl. 8, No. 15. Inscr. 5 li. N. P.
25	65	Kadashman-EN-LIL.	Fragm. of an agate cameo, d. 3.6. *Nippur* III, same place as Pl. 8, No. 15. Inscr. 5 ll. C. B. M. 8674.
25	66	[Kadashman]-EN-LIL.	Fragm. of an agate ring, orig. d. 2.7 (of the hole, 0.9), w. 0.96. *Nippur* III, same place as Pl. 8, No. 15. C. B. M. 8675.
25	67	[Kadashman ?]-Buriash.	Fragm. of an agate ring, Rev. of No. 66.
25	68	[Kadashman ?-Bu]riash.	Irregular block of lapis lazuli, convex on the inscribed surface, 13 × 7.35 × 3. *Nippur* III, same place as Pl. 8, No. 15. Inscr. 11.5 × 5.9, 3 col., 63 li. (orig. 69 ?). Sq.
25	69	Shagashalti-Shuriash.	Magnesite knob of a sceptre (or conventionalized form of a phallus), top rounded, base flat, round hole in the centre, h. c. 5, d. 7. *Nippur* III, same place as Pl. 8, No. 15. Inscr. around the top. N. P.
26	70	Bibeiashu.	Magnesite knob of a sceptre (or conventionalized form of a phallus), top rounded, base flat, round hole in the centre, h. 4.6, d. 6.8. *Nippur* III, same place as Pl. 8, No. 15. Inscr. around the top. C. B. M. 8729.

PLATE.	TEXT.	DATE.

PLATE.	TEXT.	DATE.	DESCRIPTION.
26	71	Bibeiashu.	Lapis lazuli tablet, 2.35 × 2.16. *Nippur* III, same place as Pl. 8, No. 15. Inscr. 5 li. C. B. M. 8682.
26	72	[Bibeia-]shu.	Fragm. of an axe in imitation of lapis lazuli, 11 × 6.95 × 1.25. *Nippur* III, same place as Pl. 8, No. 15. Inscr. 3 li. C. B. M. 8680.
26	73	Cassite Dyn.	Agate cameo, d. c. 1.8. *Nippur* III, same place as Pl. 8, No. 15. C. B. M. 8683.
26	74	Kurigalzu.	Fragm. of an agate ring, d. 1, w. 1.1. *Nippur* III, same place as Pl. 8, No. 15. Inscr. 3 li. C. B. M. 8684. The ring originally formed the centre part of a votive cylinder. Cf. Pl. 22, No. 50.
26	75	Cassite Dyn.	Fragm. of an axe in imitation of lapis lazuli, 6 × 2.5 × 1.5. *Nippur* III, same place as Pl. 8, No. 15. Inscr. 6 li. C. B. M. 8681.
26	76	. . . . . . ia-ash.	Fragm. of an axe in imitation of lapis lazuli, 5.26 × 2.1. *Nippur* III, same place as Pl. 8, No. 15. Inscr. 4 li. N. P.
27	77	Cassite Dyn.	Fragment of a vase in soapstone rock, 8.5 × 8.8 (orig. d. at the bottom 13.2). *Nippur* V, c. 3 m. below the surface. Inscr. 7 li. C. B. M. 8690.
27	78	Nazi-Maruttash.	Fragm. of an axe in imitation of lapis lazuli, 6.2 × 6.2 × 1.7. *Nippur* III, same place as Pl. 8, No. 15. Inscr. 9 li. C. B. M. 8685.
27	79	[Bibeia-]shu.	Fragm. of an axe in imitation of lapis lazuli, 2.35 × 2.85 × 1.5. *Nippur* III, same place as Pl. 8, No. 15. Inscr. 4 li. C. B. M. 8686.
27	80	c. 1100 B.C.	Fragm. of a reddish granite (boundary) stone of phallic shape, l. 15.5. *Nippur* III, c. 1.5 m. below the surface on the slope of the T. hill N.W. of Z. Inscr. 2 col., 8 li. Pho. and N. P. Cf. Pl. XII, 32, 33.
28	81	Rammân-shum-usur.	Fragm. of a baked brick, yellowish, very soft, partly covered with bitumen, 22.5 (fragm.) × 18.4 (fragm.) × 6.9 (orig.). *Nippur* III, found out of place in a later structure of the inner wall of Z. (cf. Pl. 29, No. 82; Pl. 10, No. 18; Pl. 13, No. 22; Pl. 20, No. 38). Inscr. written, 15.2 × 8.6, 10 li. C. B. M. 8643.
29	82	Mili-Shikhu.	Brick of baked clay, brown, partly covered with bitumen, 29.6 × 30.2 × 6.7. *Nippur* III, inner wall of Z. Every brick of this structure bears the name of Mili-Shikhu with exactly the same inscription (stamped), except a few which belong to Ur-Ninib (Pl. 10, No. 18), Bur-Sin (Pl. 11, No. 19), Kurigalzu (Pl. 20, No. 38), Rammânshumusur (Pl. 28, No. 81). The latter four evidently formed a part of the ancient structure, and were utilized by Mili-Shikhu in his restoration of the platform of Z. Inscr. stamped, 14.8 × 7, 11 li. C. B. M. 8632. Cf. Pinches "An Early Babylonian Inscription from Niffer" in *Hebraica* VI, pp. 55–58.
30	83	Bêl-nâdin-aplu.	Black limestone tablet, 16.75 × 12.1 × 5.1. Presumably neighborhood of *Babylon*. Obv., slightly rounded, 22 li. C. B. M. 13.
31	83	Bêl-nâdin-aplu.	The same, Rev., rounded, 24 li.
32	84	Nabopolassar.	Cylinder of baked clay, cartridge shaped, hollow, small hole at the top, dark brown with grayish spots; when found, half covered with bitumen; h. 15.2, d. of the base 8.85, d. of the hole 2.2. *Babylon.* Inscr. 3 col., 45 + 65 + 59 = 169 li. C. B. M. 9090. Cf. Pl. XIII, No. 34. The variants have been taken from a mutilated cylinder (B) in the British Museum, published by Strassmaier in *Z. A.* IV, pp. 129–136. Apparent mistakes in Strassmaier's edition

PLATE.	TEXT.	DATE.	DESCRIPTION.
			are not quoted as variants (cf. also Strassmaier in *Z. A.* IV, pp. 106–113, and Winckler in Schrader's *Keilinschriftliche Bibliothek* III, Part 2, pp. 2-7).
33	84	Nabopolassar.	The same, continued.
34	85	Nebuchadrezzar II.	Fragm. of a baked clay cylinder, barrel shaped, solid, light brown; h. 23.9, d. 8.8 at the top and base, 11.5 at the centre. *Babylon.* Inscr. 4 col., 23 (orig. c. 48) + 32 (orig. c. 56) + 30 (orig. c. 56) + 28 (orig. c. 48) = 113 (orig. c. 208) li. C. B. M. 1785. Cf. Pl. XIV, No. 35. According to information of the Arabs the cylinder was found whole and intentionally broken lengthwise. The other half is supposed to be in existence.
35	85	Nebuchadrezzar II.	The same, columns III, IV.

II. PHOTOGRAPH (HALF-TONE) REPRODUCTIONS.

PLATE.	TEXT.	DATE.	DESCRIPTION.
I	1	Sargon I.	Door socket in diorite. *Nippur.* Cf. Pl. 1.
II	2	Sargon I.	Brick stamp of baked clay, Rev. *Nippur.* Cf. Pl. 3, No. 3.
II	3	Narâm-Sin.	Brick stamp of baked clay, Obv. *Nippur.* Cf. Pl. 3, No. 4.
III	4–12	Âl-usharshid.	Fragments of vases from which the text on Pl. 4 has been obtained. *Nippur.* Nos. 4, 5: dolomite; Nos. 6, 8, 9, 10: white marble; No. 7: red banded marble of agate structure; Nos. 11, 12: white marble of stalactitic structure. For the restoration of li. 6 fragm. 8860 (white marble) has been consulted.
IV	13	Âl-usharshid.	Fragm. of a vase in reddish numulite limestone. *Nippur.* Cf. Pl. 5, No. 6.
V	14	Âl-usharshid.	Fragm. of a white marble vase with gray and reddish veins here and there. *Nippur.* Cf. Pl. 5, No. 7.
VI	15	Not later than 2400 B.C.	Fragm. of a white marble slab, 26.65 × 15.8 × 7.9. *Abu Habba.* Orig. inal in Constantinóple. Photograph taken from a cast. Inscr. on both sides and left edge, 391 li. Obv., 9 col., (20 + 25 + 24 + 22 + 22 + 26 + 19 + 23 + 4 =) 185 li.
VII	16	Not later than 2400 B.C.	The same, Rev., 9 col., (19 + 19 + 23 + 25 + 28 + 24 + 25 + 22 + 13 =) 198 li.
VIII	17	Not later than 2400 B.C.	The same, left edge, 1 col., 18 li.
VIII	18, 19	c. 2400 B.C.	Tablets of baked clay, reddish brown with black spots. These tablets have a peculiar shape; they are rounded at both ends and on the left side, but angular and flat on the right side, as if cut off from a larger tablet. *Yokha.* No. 18 : 10.3 × 4.3, th. 1.6 on the left, 2.2 on the right side. C. B. M. 9042. No. 19 : 10.62 × 4.5, th. 1.7 on the left, 2.55 on the right side. C. B. M. 9041.
IX	20	Hammurabi.	Fragm. of an ornamented stamp in the shape of a vase, made of soapstone (composed of a green micaceous and very soft mineral, probably talc). Presumably neighborhood of *Babylon.* Cf. Pl. 15, No. 27.

PLATE.	TEXT.	DATE.	DESCRIPTION.
IX	21	Burna-Buriash.	Fragm. of a white marble mortar. Presumably neighborhood of Babylon. Cf. Plates 16, 17.
X	23	Burna-Buriash.	Knob of a sceptre (or conventionalized form of a phallus) in ivory. Side view. Nippur. Cf. Pl. 18, No. 34.
X	22, 24	Nazi-Maruttash.	Knobs of sceptres (cf. Pl. X, 23) in magnesite. Top views. Nippur. Cf. Pl. 23, Nos. 57, 56.
XI	25	Kurigalzu.	Inscribed block of lapis lazuli, tablet in process of cutting. Nippur. Cf. Pl. 18, No. 36.
XI	26	Kurigalzu.	Fragm. of a votive battle axe in imitation of lapis lazuli (blue glass). Nippur. Cf. Pl. 20, No. 39.
XI	27	c. 1350 B.C.	Fragm. of a votive battle axe in imitation of lapis lazuli, 8.32 × 5.65 × 5.1. Nippur III, same place as Pl. 8, No. 15. C. B. M. 8800.
XI	28	c. 1350 B.C.	Fragm. of a votive battle axe in lapis lazuli, 6.4 × 5.7 × 1.5. The inscription has been erased in order to use the material. Nippur III, same place as Pl. 8, No. 15. C. B. M. 8597.
XII	29–31	c. 1150 B.C.	Three small fragments of an inscribed bas-relief in a basaltic stone, h. c.5. Nippur III, on the S.E. side of the Bur-Sin shrine (cf. Pl. 11, No. 19).
XII	32, 33	c. 1100 B.C.	Fragm. of a reddish granite (boundary) stone of phallic shape. Nippur. Two views of the same stone. Cf. Pl. 27, No. 80.
XIII	34	Nabopolassar.	Cylinder of baked clay, cartridge-shaped, hollow, small hole at the top. Babylon. Cf. Plates 32, 33.
XIV	35	Nebuchadrezzar II.	Cylinder of baked clay, barrel-shaped, solid. Babylon. Cf. Plates 34, 35.
XV	36	1889 A.D.	Plan of the first year's excavations at Nippur (February 5 to April 16).

CUNEIFORM

TEXTS.

Pl. 1

1

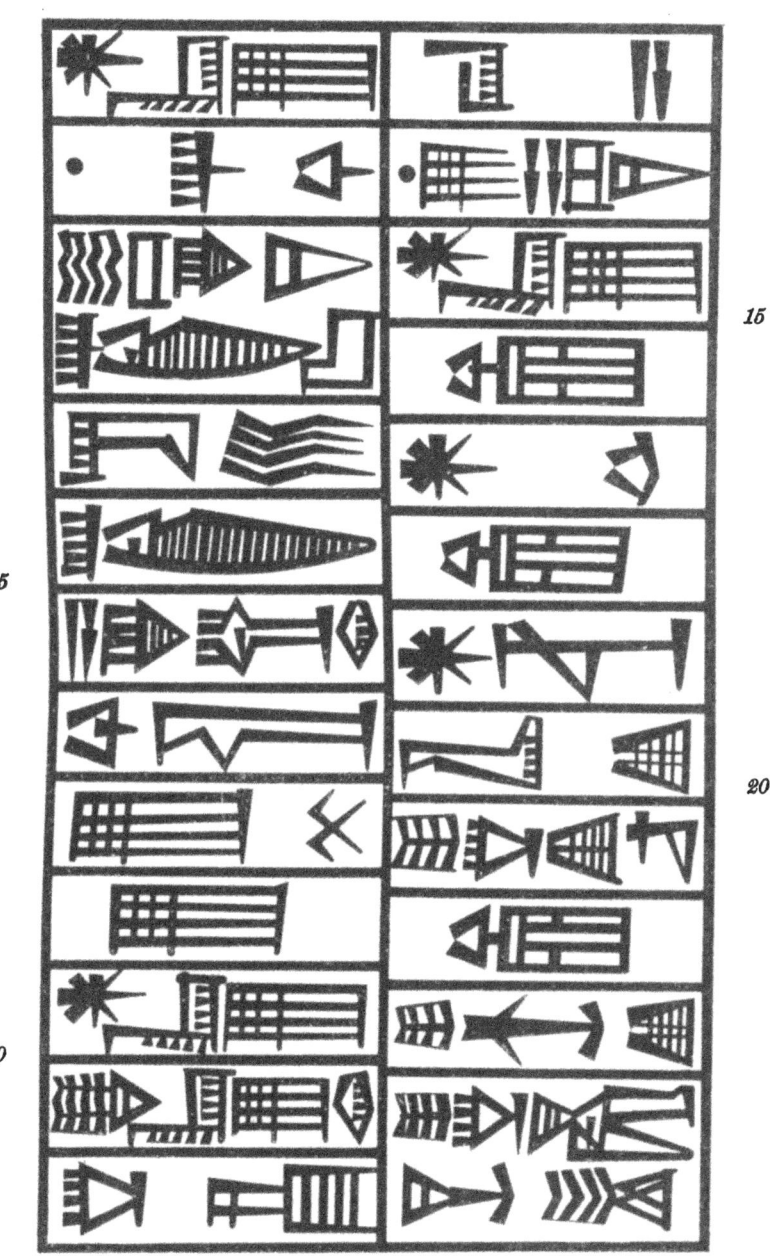

Pl. 2

2

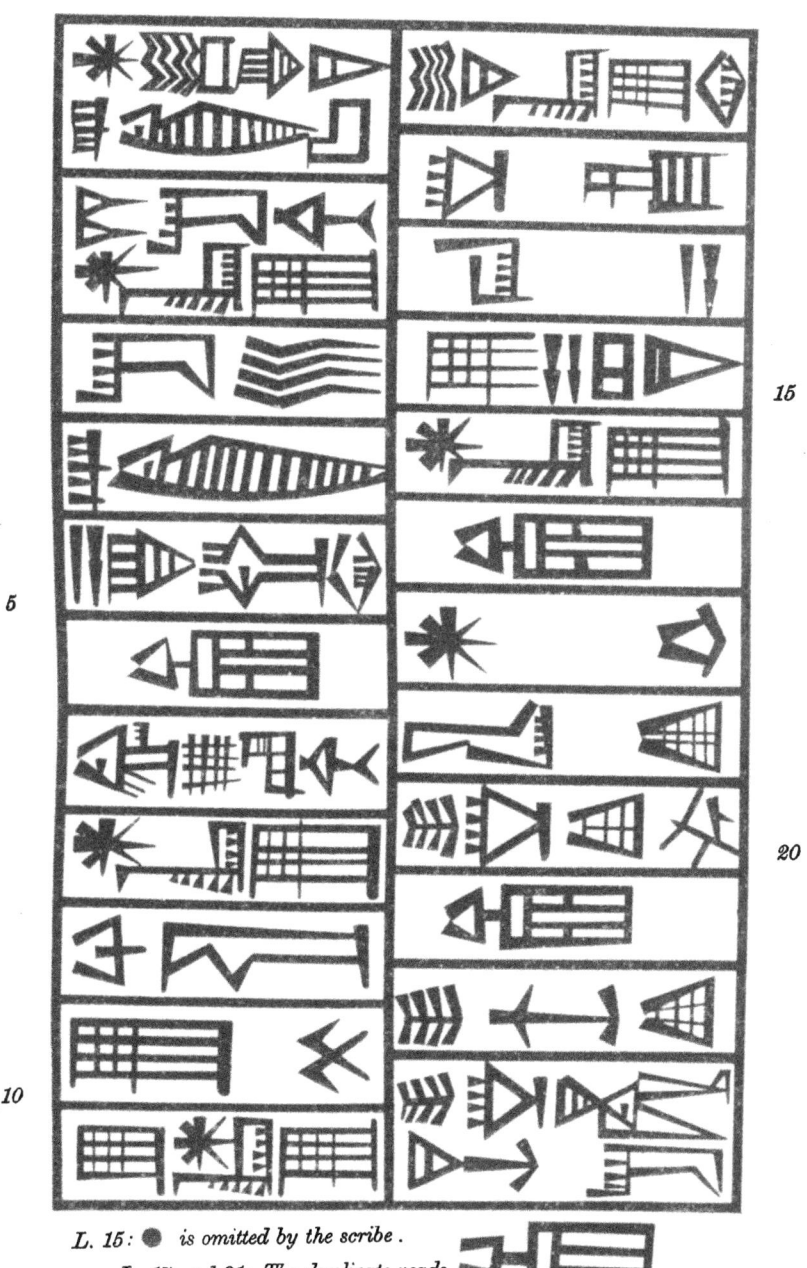

5

10

15

20

L. 15: ● is omitted by the scribe.
L. 17 and 21: The duplicate reads

Pl. 3

3

4

Pl. 4

10

Pl. 5

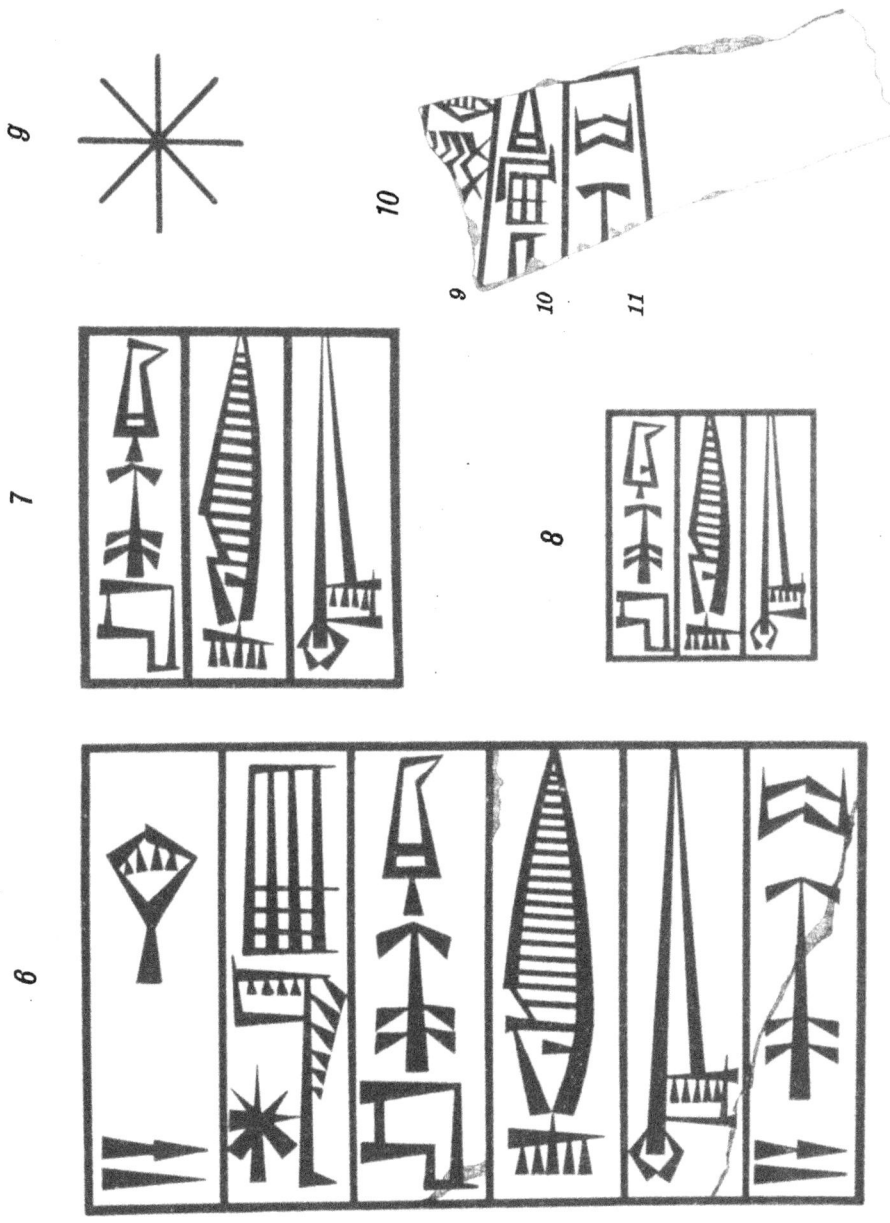

Pl. 6

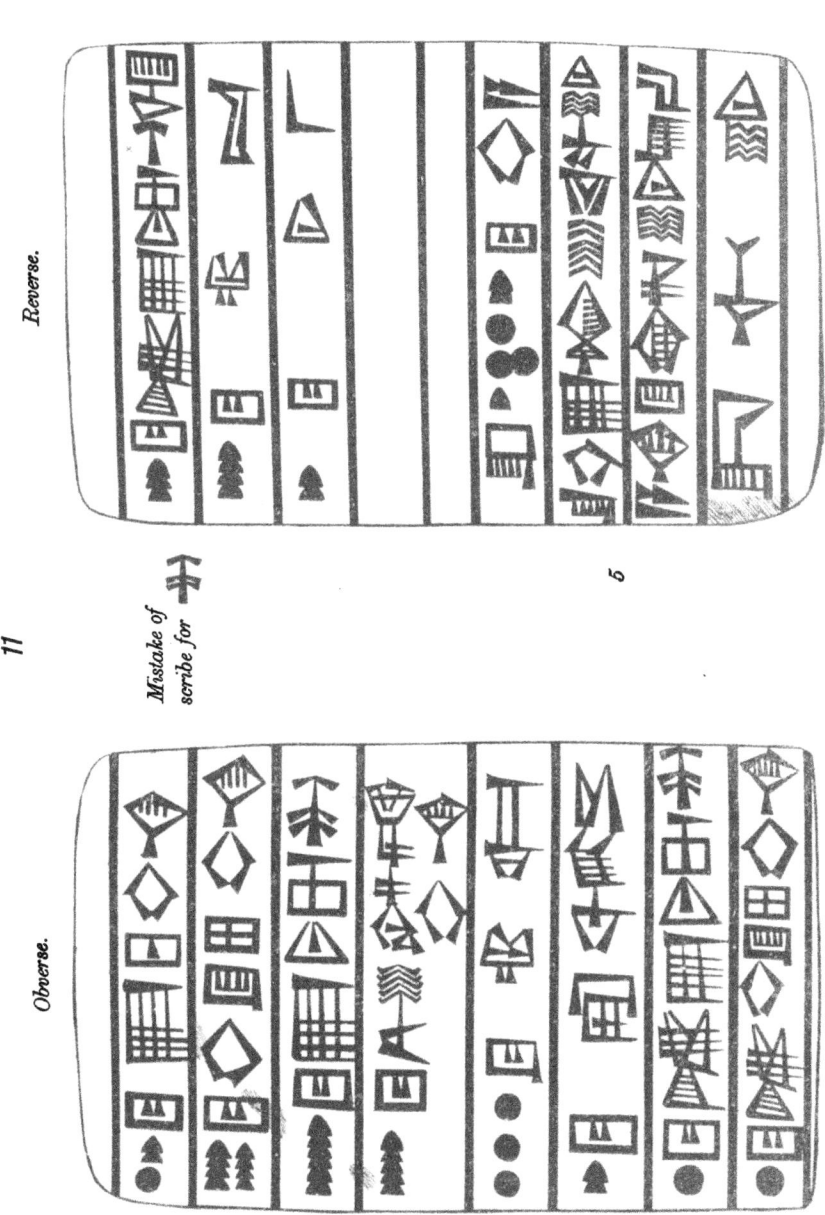

Reverse.

Obverse.

Mistake of scribe for

11

5

6

Pl. 7

12

13

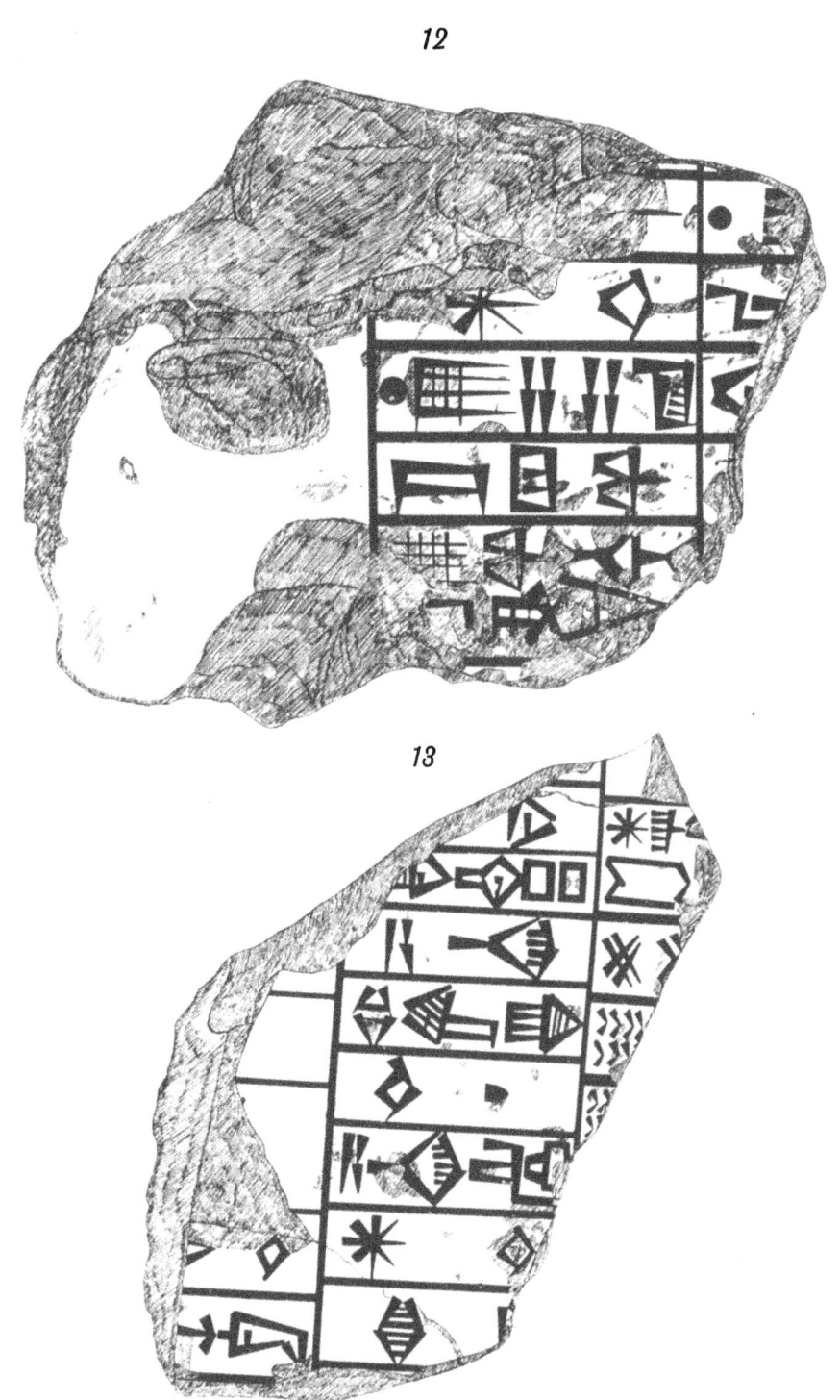

Pl. 8

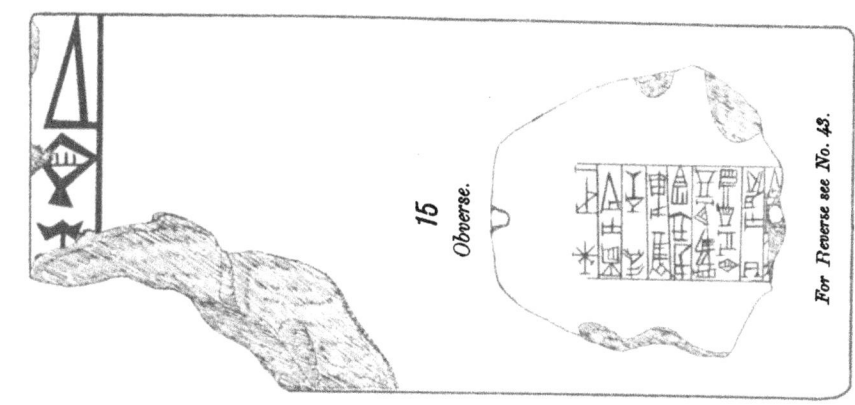

Pl. 9

16

Obverse.

Reverse.

5

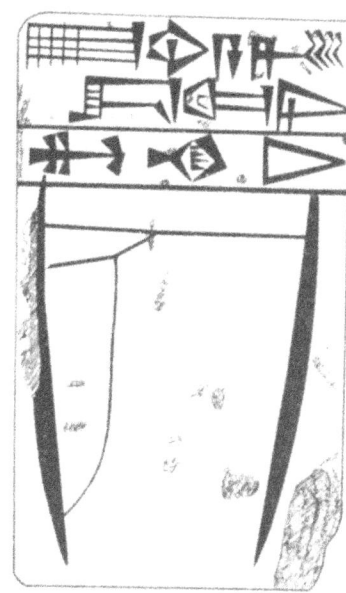

17

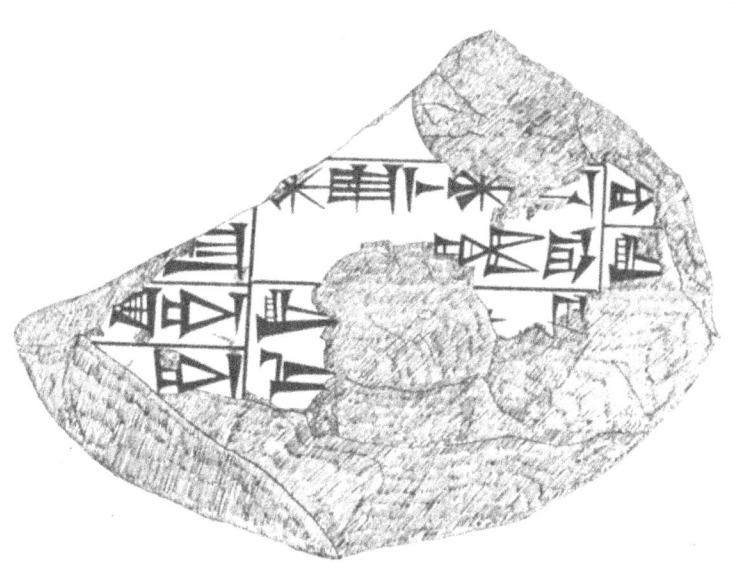

Pl. 10

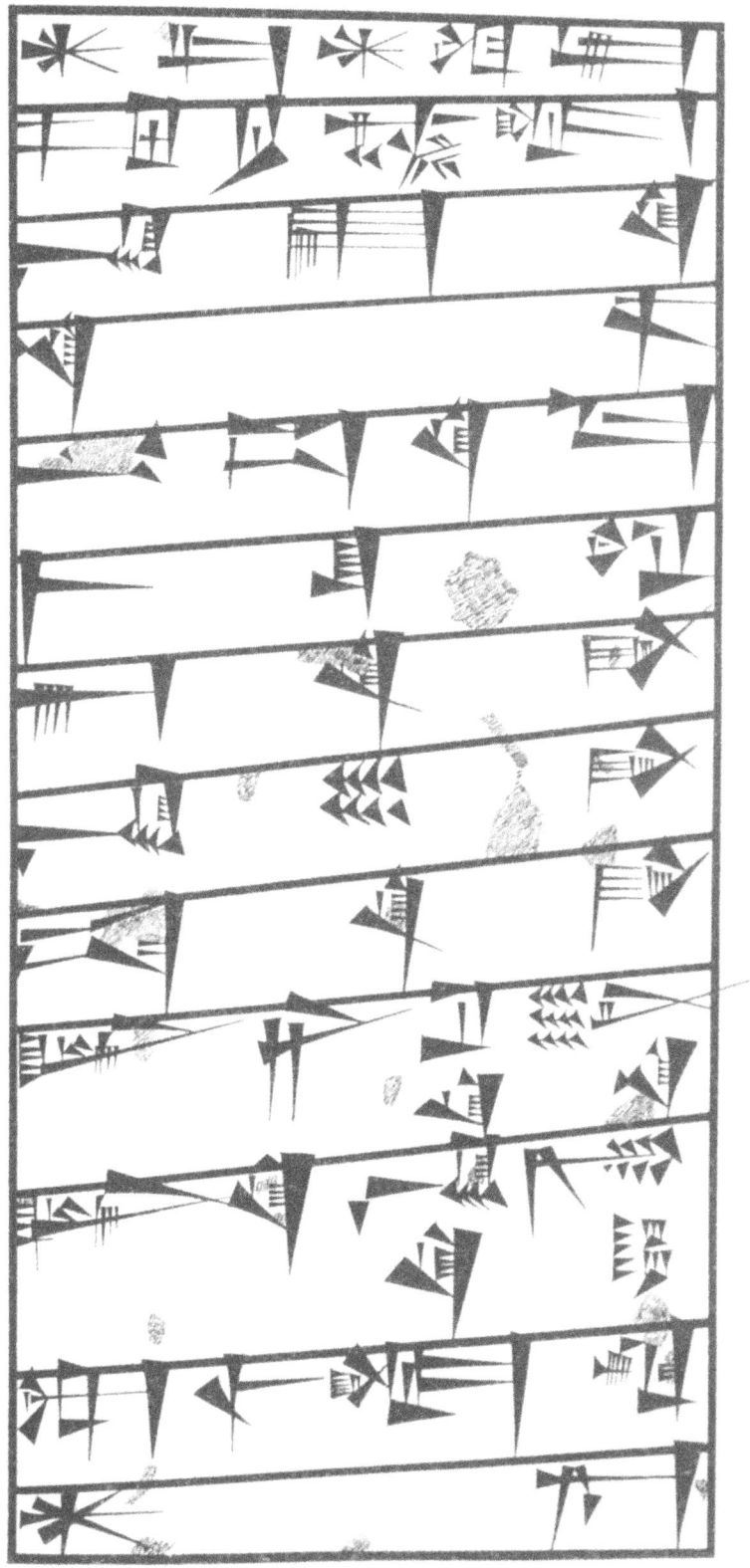

Trans. Am. Phil. Soc., N. S. XVIII, 1.

19

Pl. 11

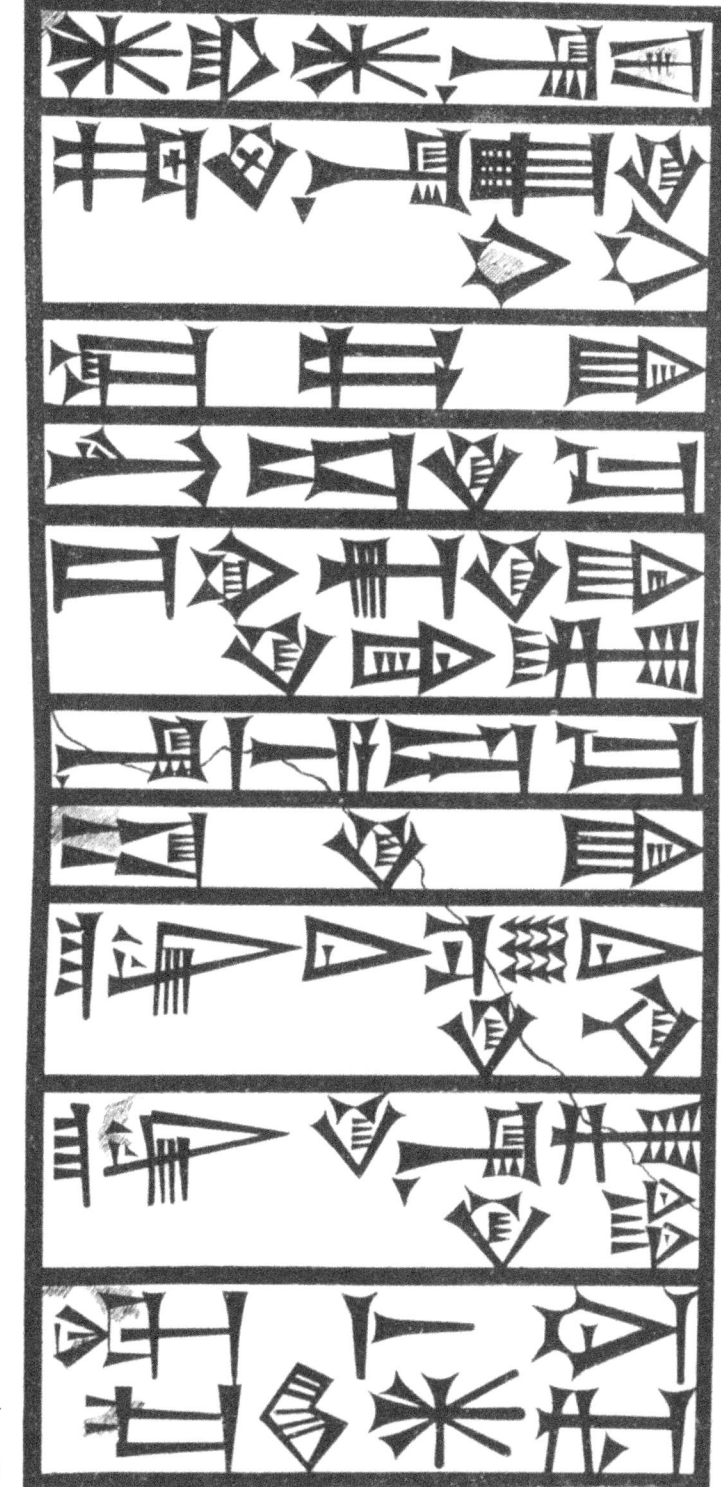

Pl. 12

20

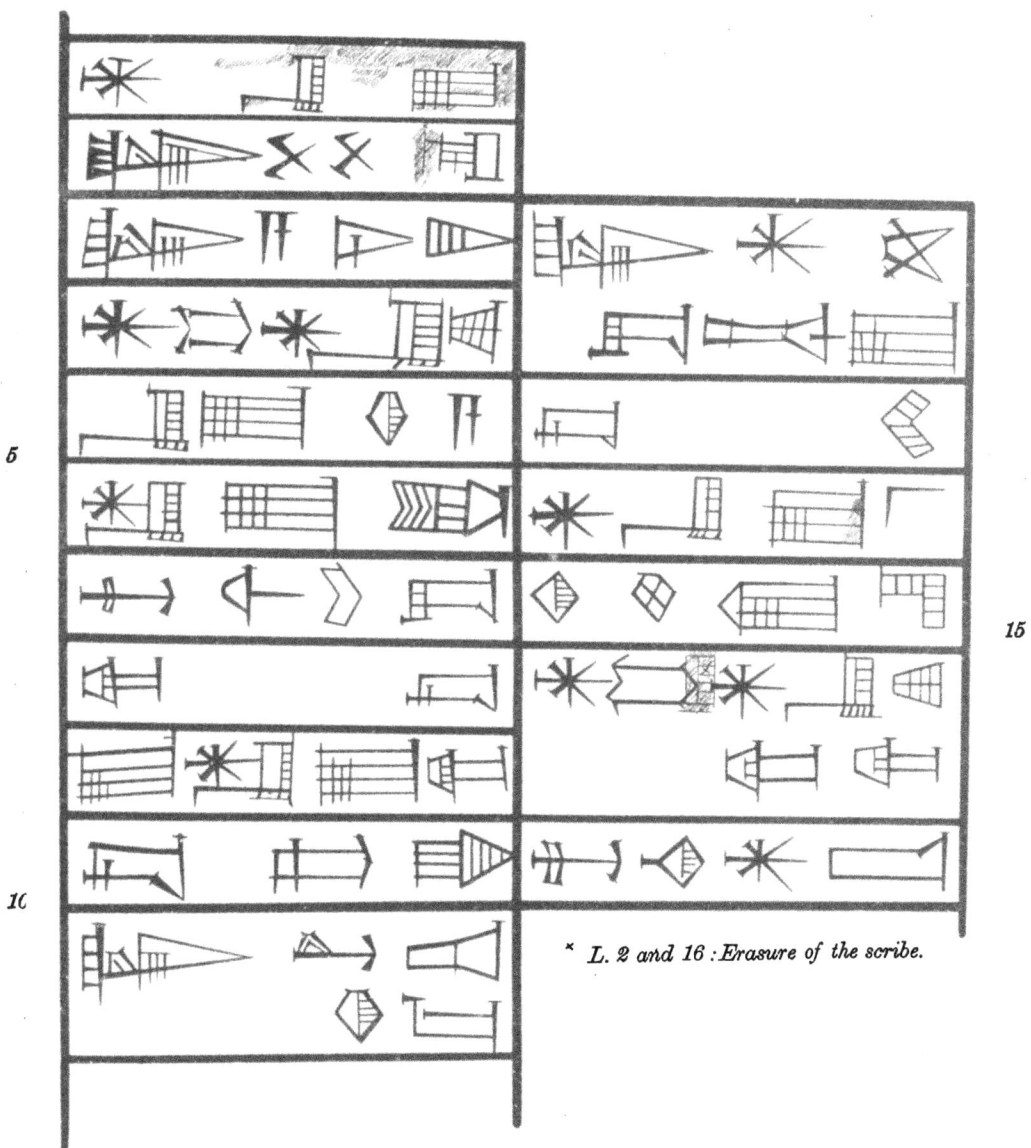

* L. 2 and 16 : Erasure of the scribe.

Pl. 13

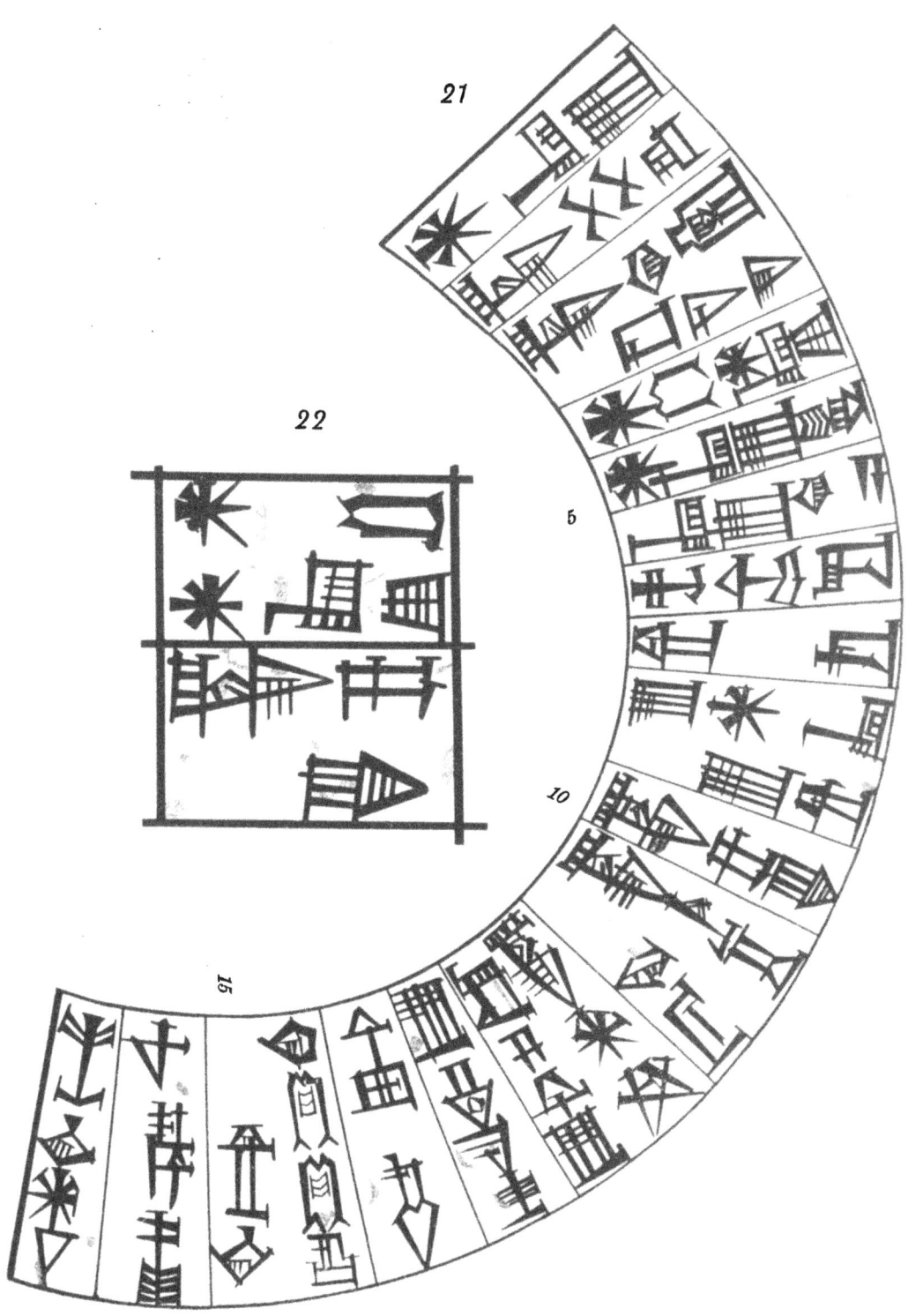

Pl. 14

23

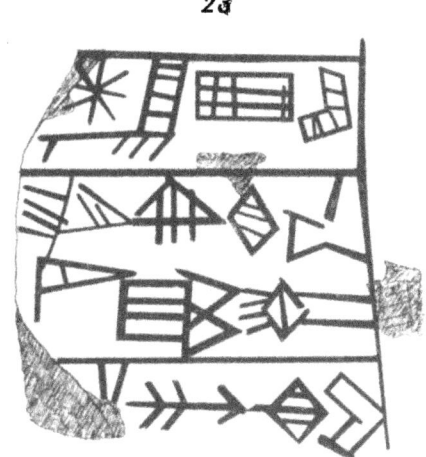

24

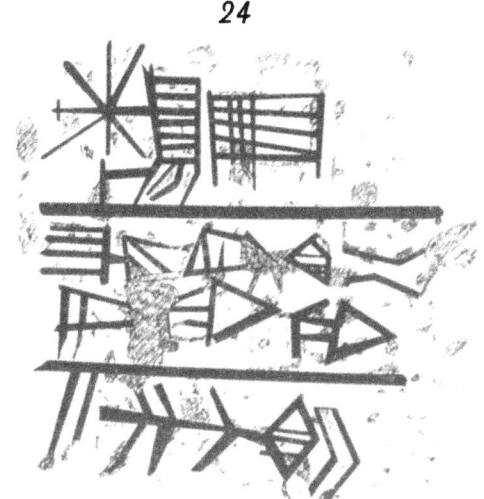

25

Trans. Am. Phil. Soc., N. S. XVIII, 1.

Pl. 15

26

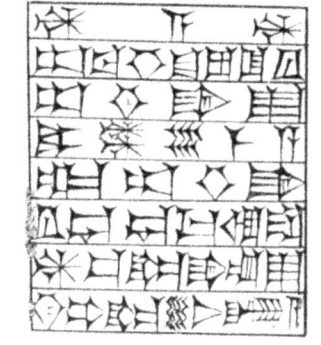

28

29

27

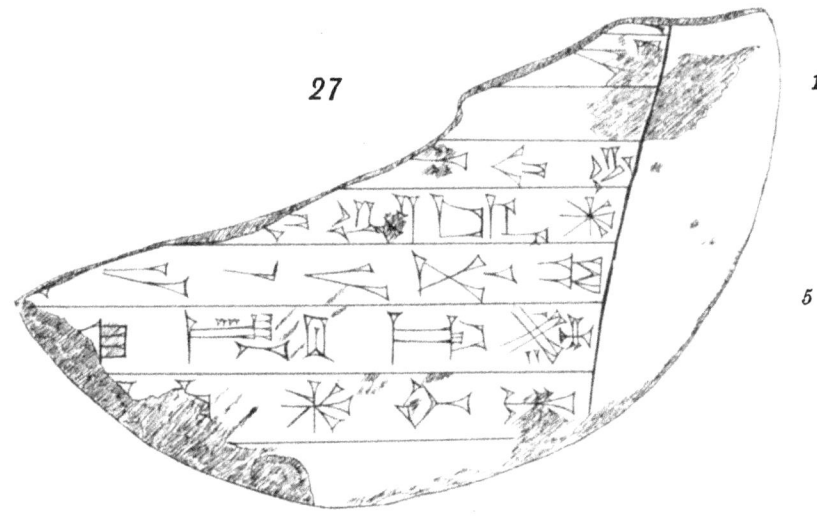

30

31

32

Pl. 16

33

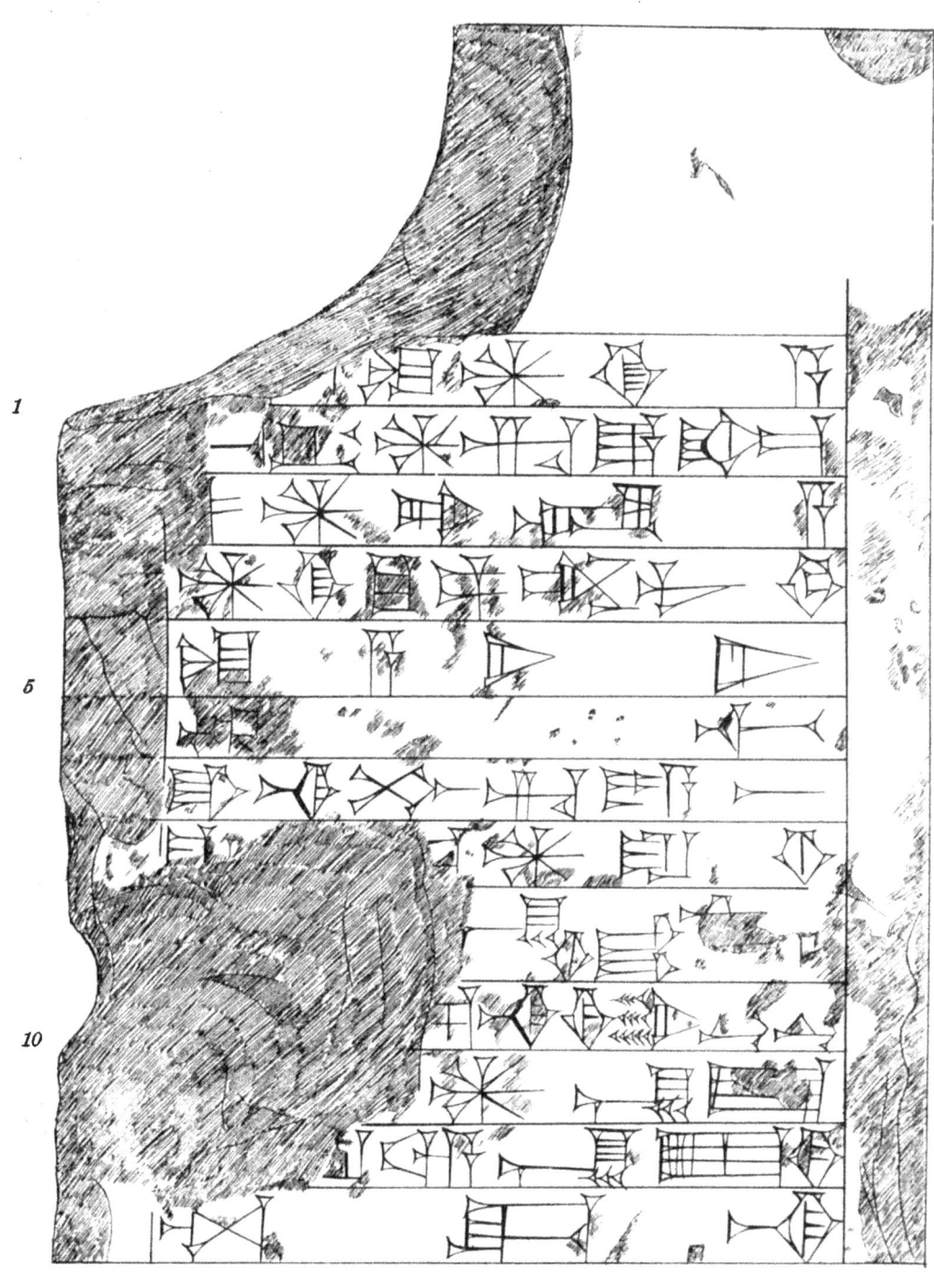

Pl. 17

33
Continued

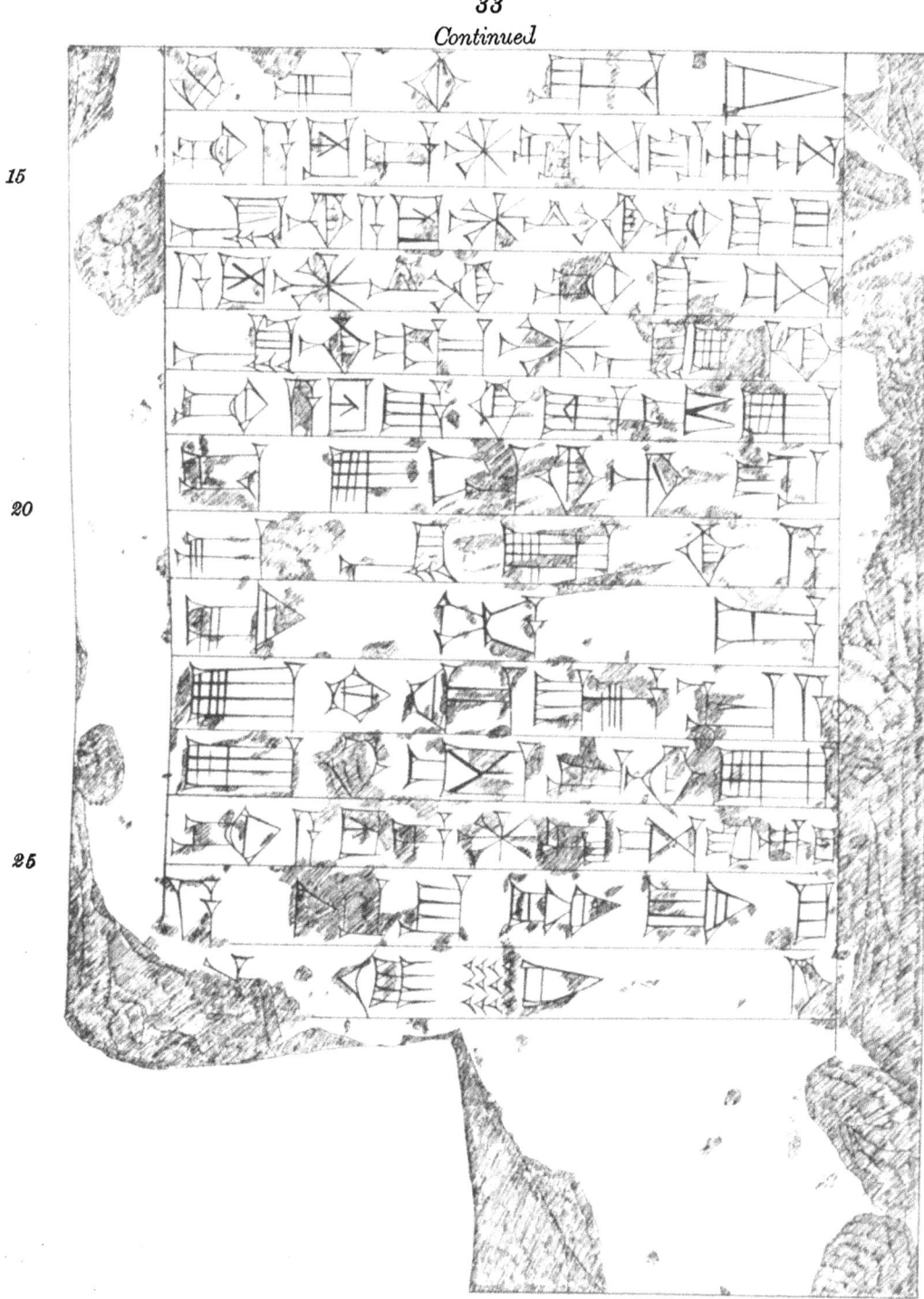

Pl. 18

34

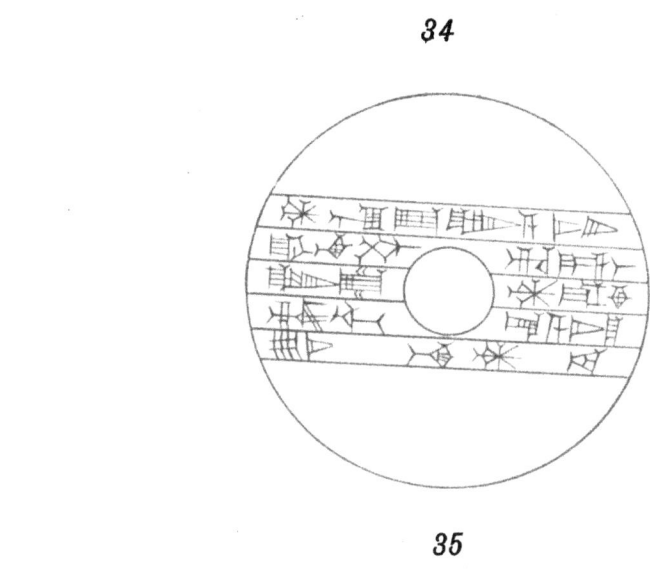

35

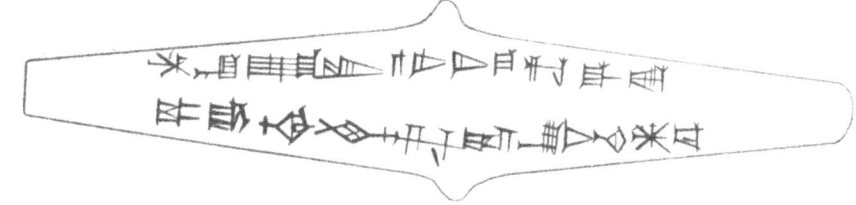

36

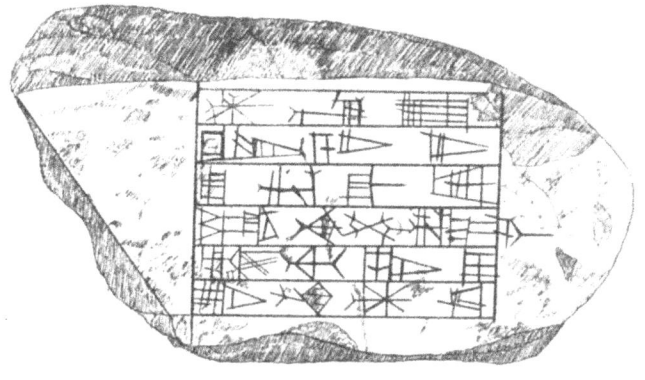

Pl. 19

37

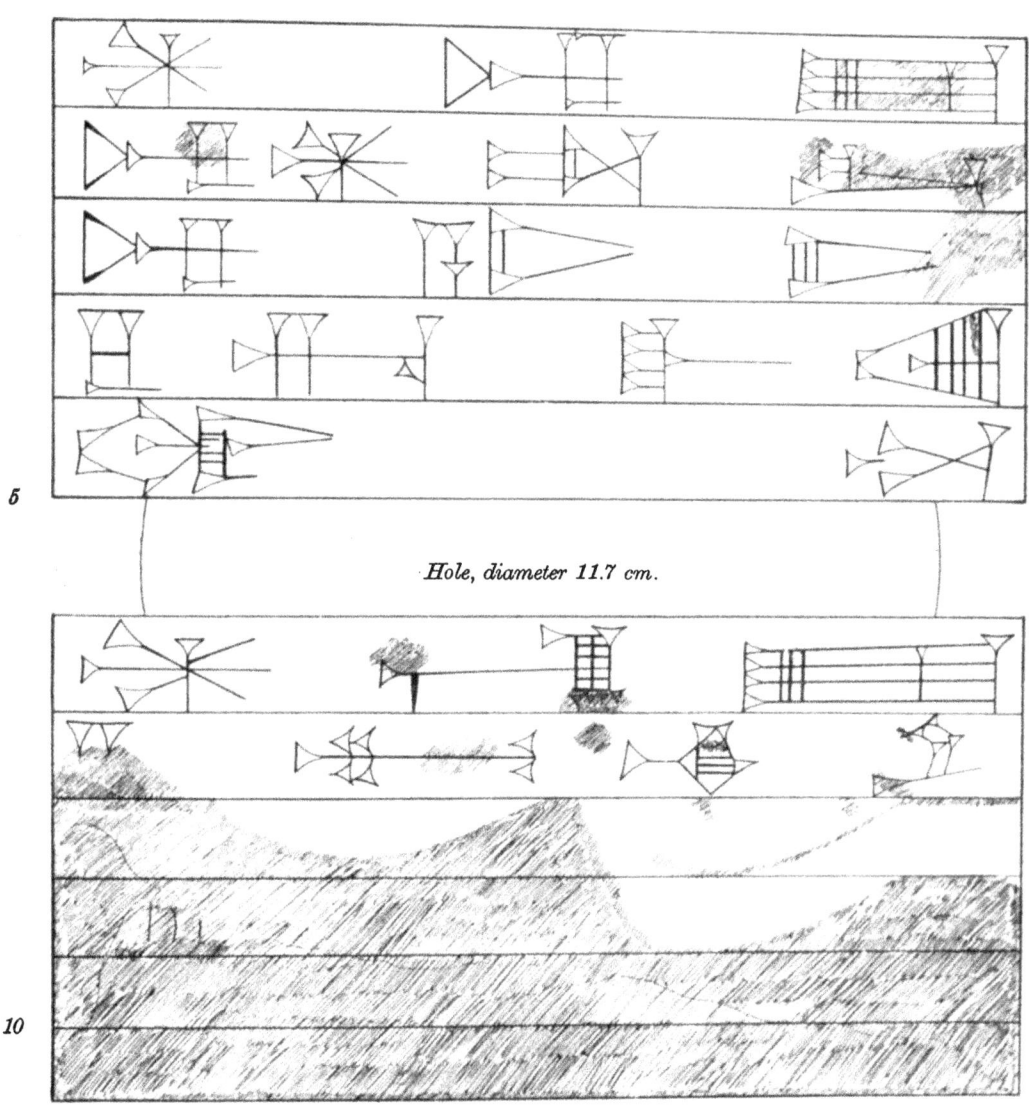

5

Hole, diameter 11.7 cm.

10

Pl. 20

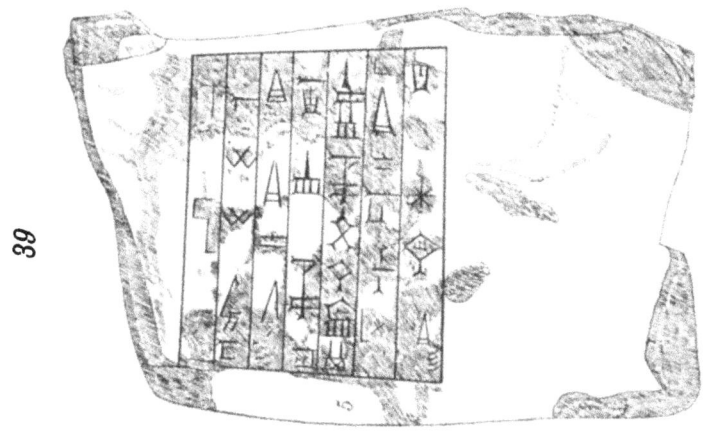

39

38

Trans. Am. Phil. Soc., N. S. XVIII, 1.

Pl. 21

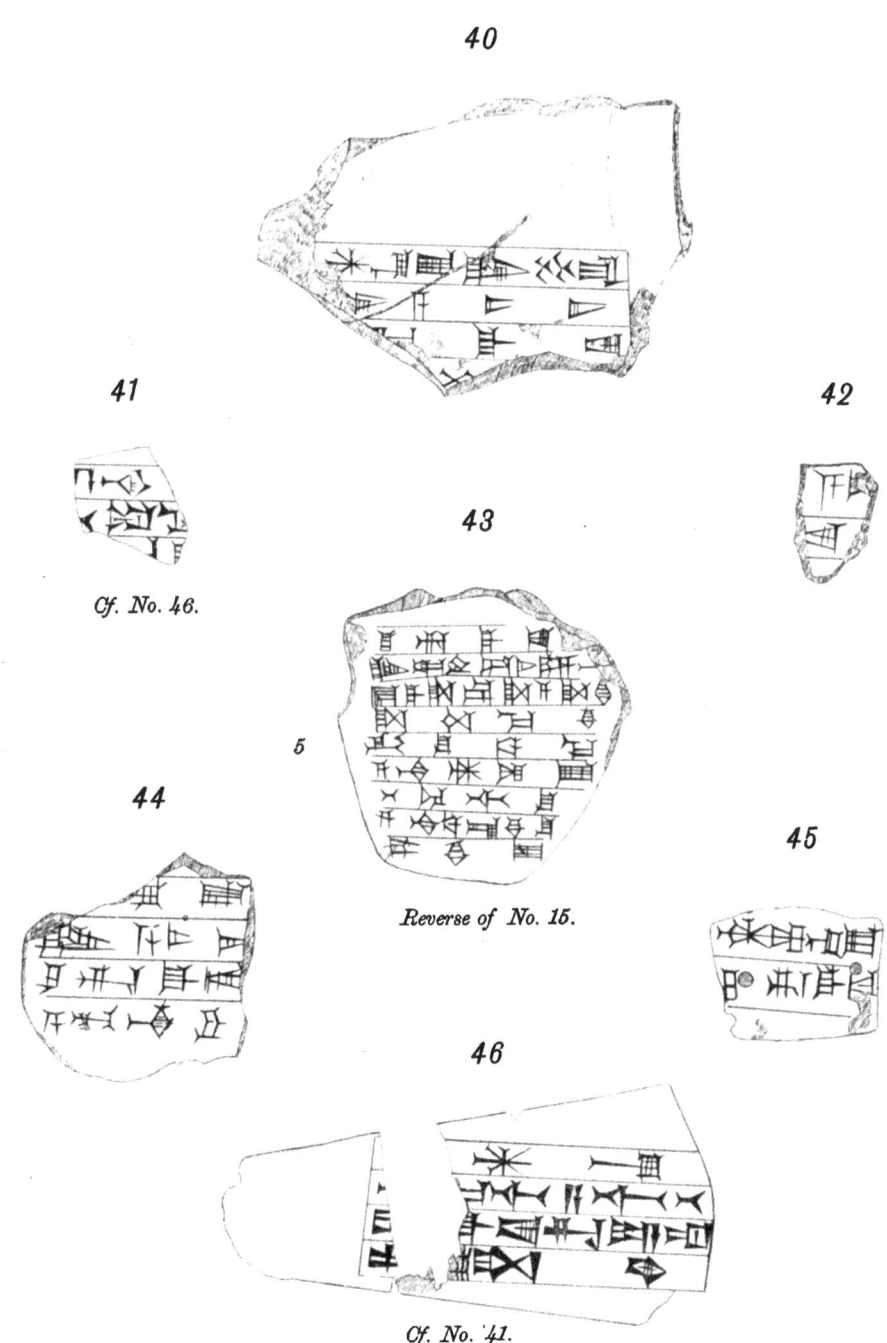

40

41

42

Cf. No. 46.

43

44

45

5

Reverse of No. 15.

46

Cf. No. 41.

Pl. 22

47

The second perpendicular
line is a mistake of the scribe

Erasure of
the scribe

5

48

5

Mistake of scribe
for

49

50

5

Cf. No. 74.

51

52

53

54

55

Pl. 23

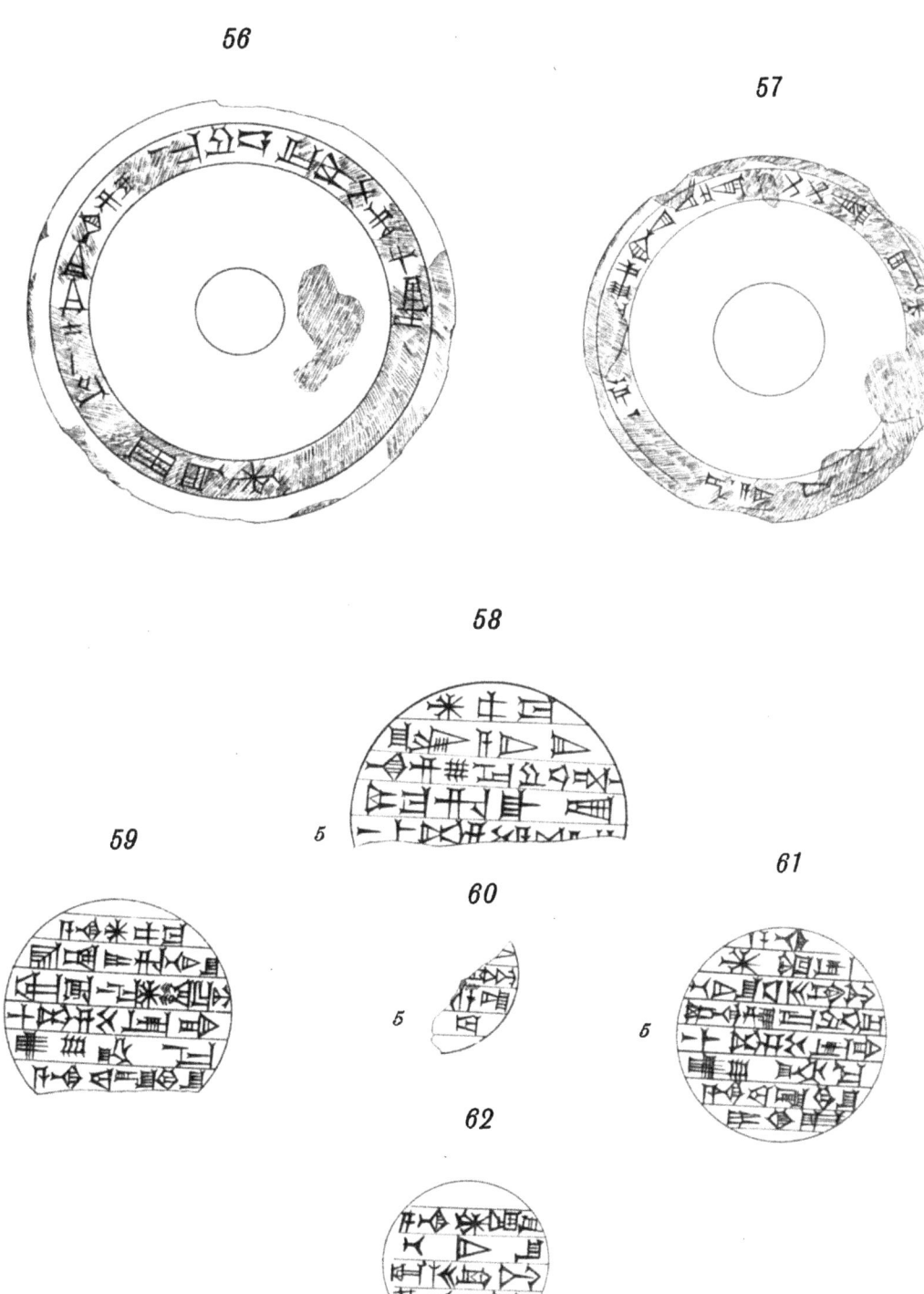

Trans. Am. Phil. Soc., N. S. XVIII, 1.

Pl. 24

63

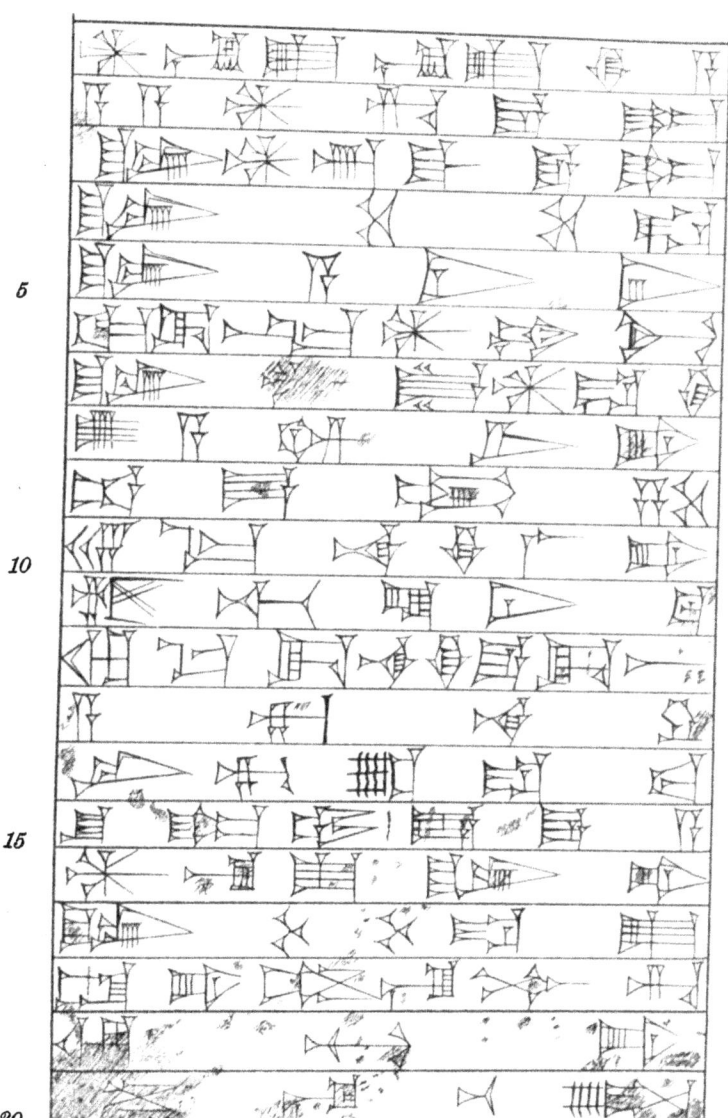

L. 7. Erasure of DINGIR, *the second character of* KA-DINGIR-RA, *written by the scribe erroneously before* KA.

Pl. 25

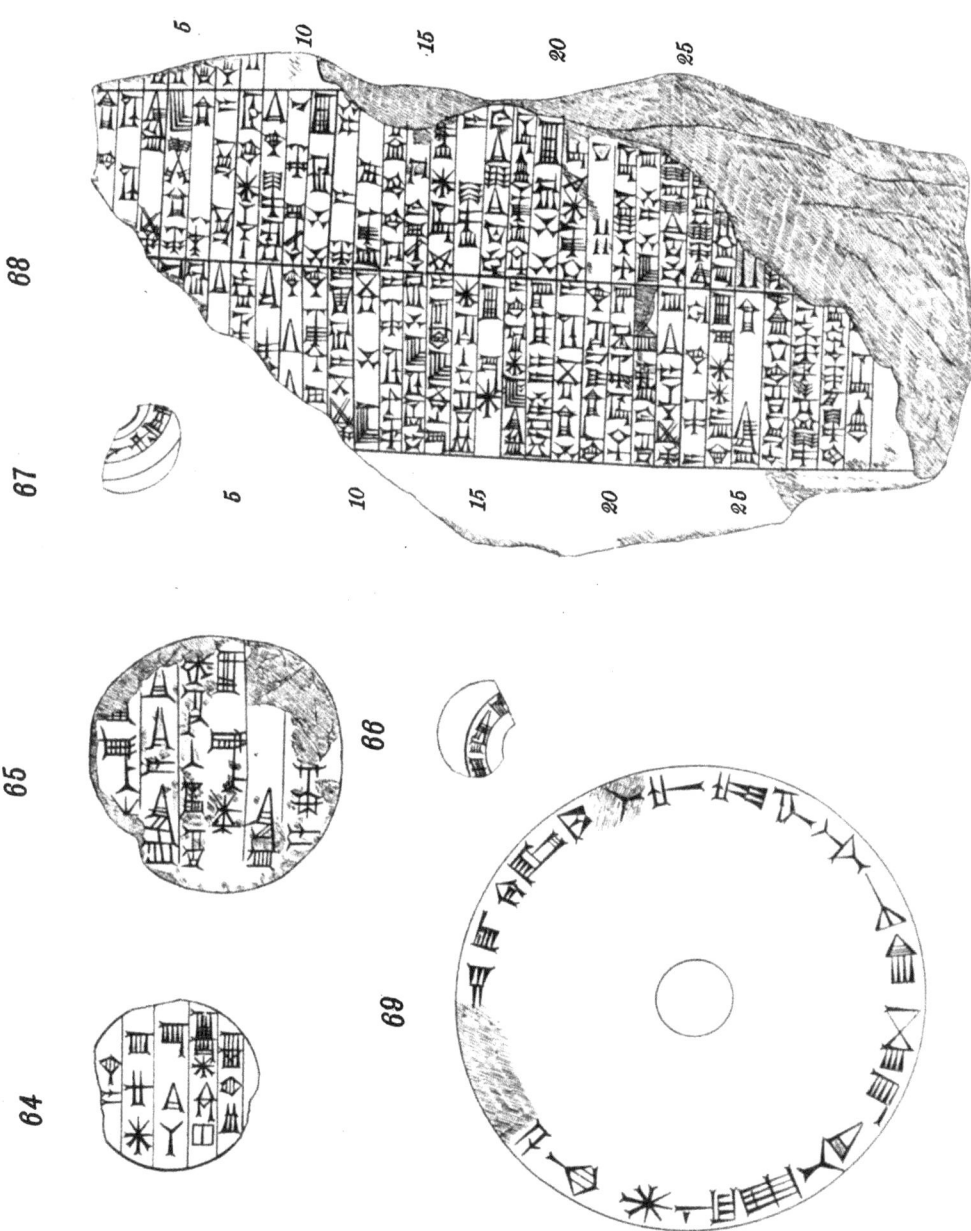

Pl. 26

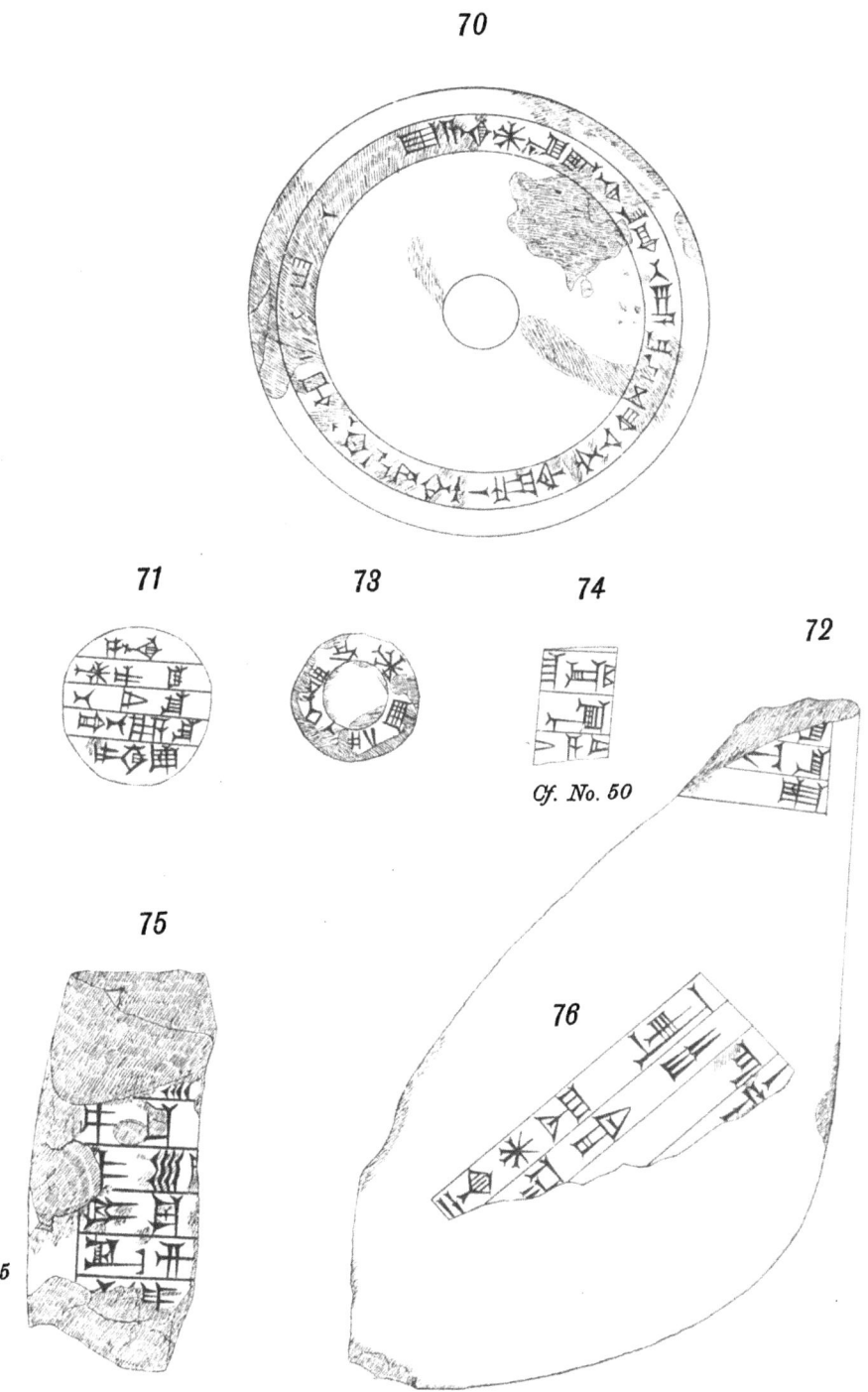

70

71   73   74   72

Cf. No. 50

75

76

5

Pl. 27

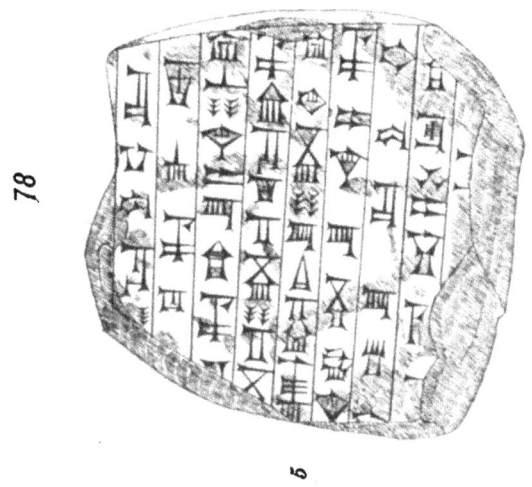

78

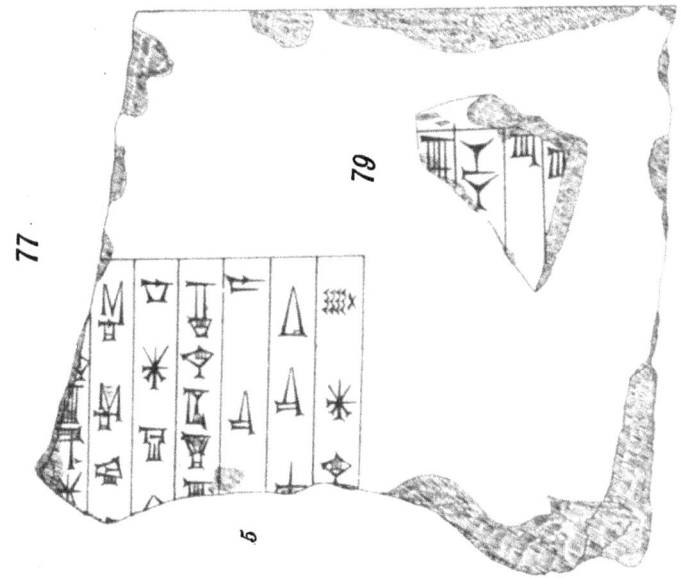

77

79

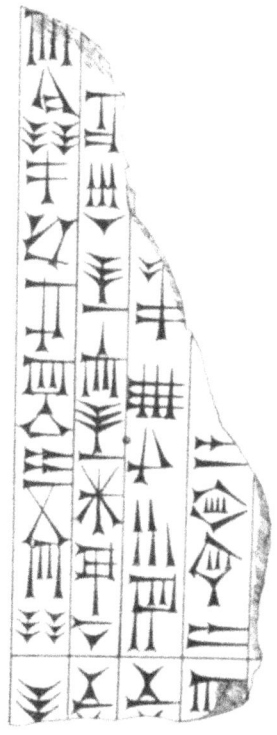

80

Pl. 28

81

5

Trans. Am. Phil. Soc., N. S. XVIII, 1.

Pl. 29

82

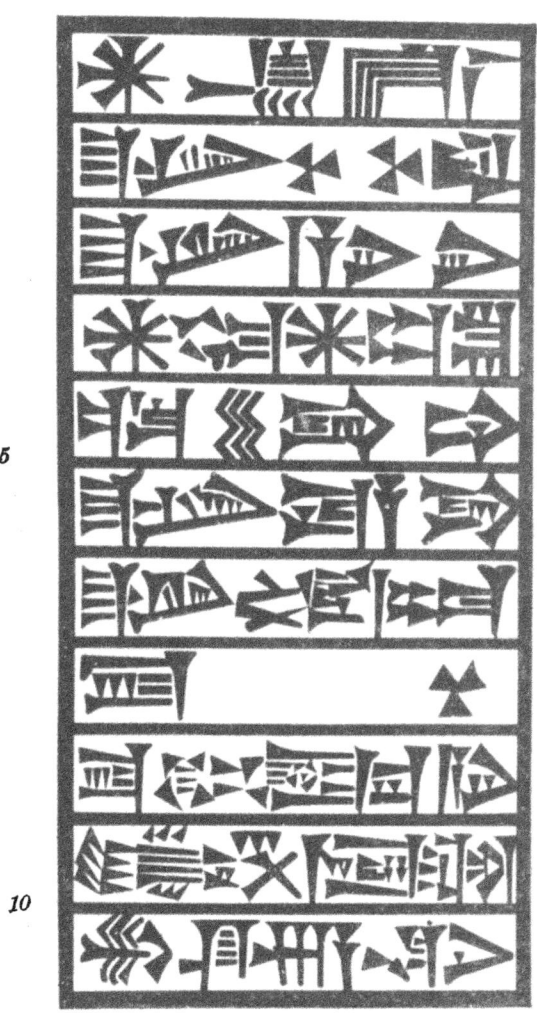

Pl. 30

83

Obverse.

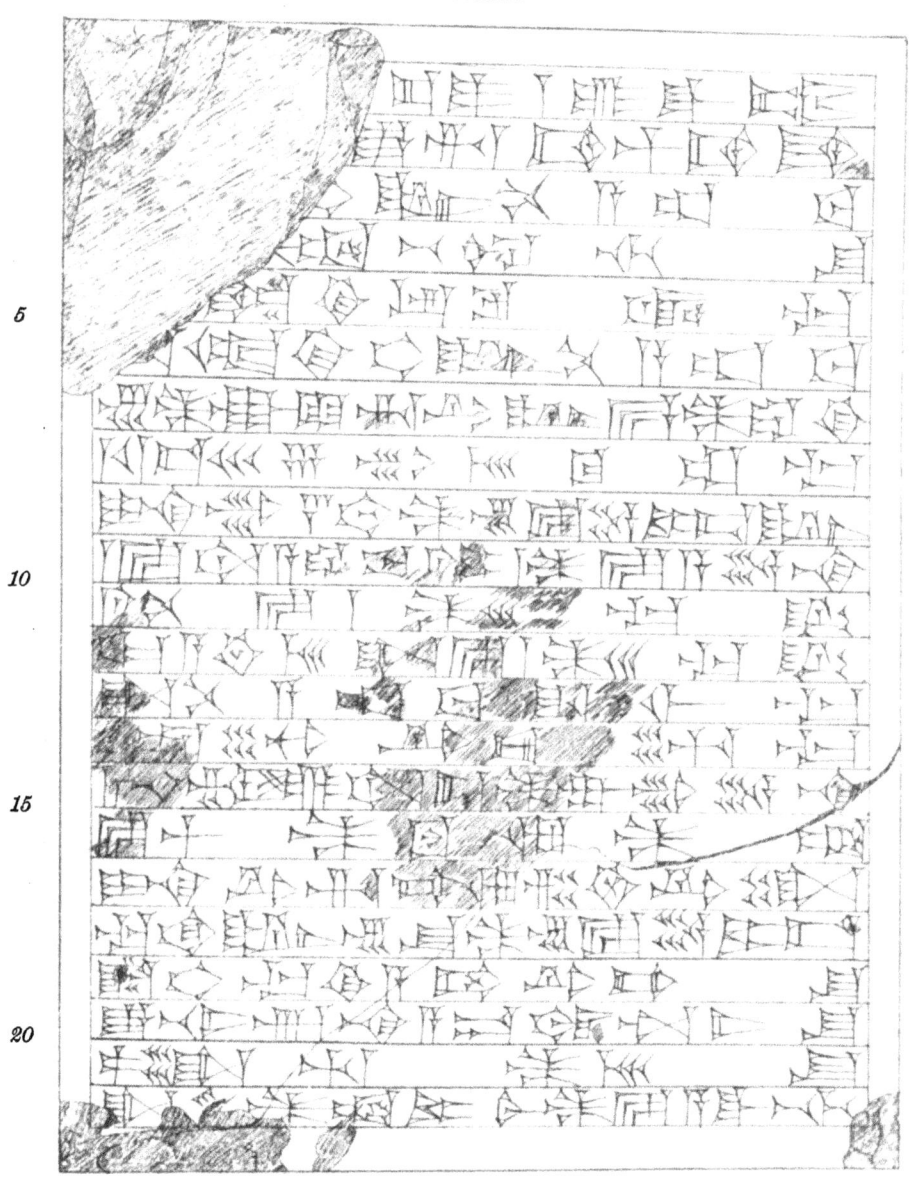

83

Reverse.

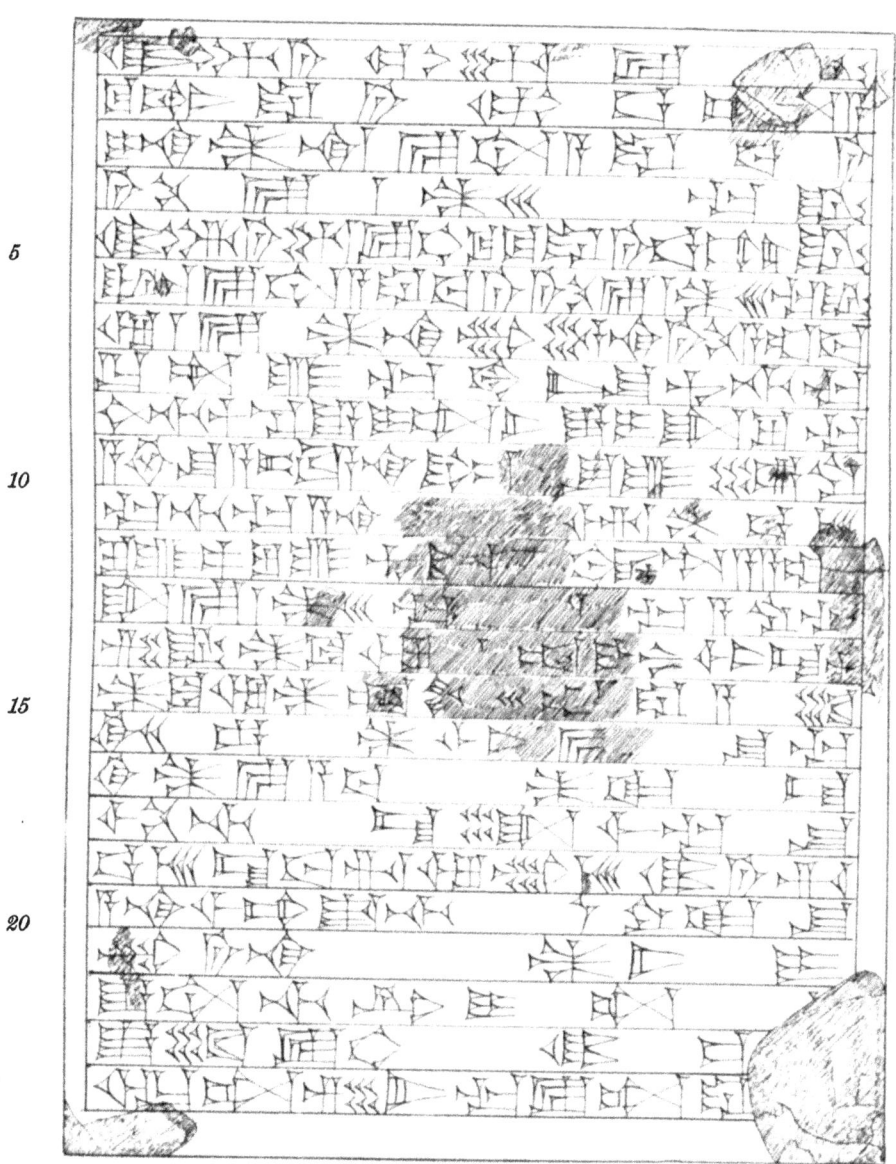

On the left margin of Reverse are traces of

On the lower margin of Reverse is

Pl. 32

Trans. Am. Phil. Soc., N. S. XVIII, 1.

84

Pl. 33

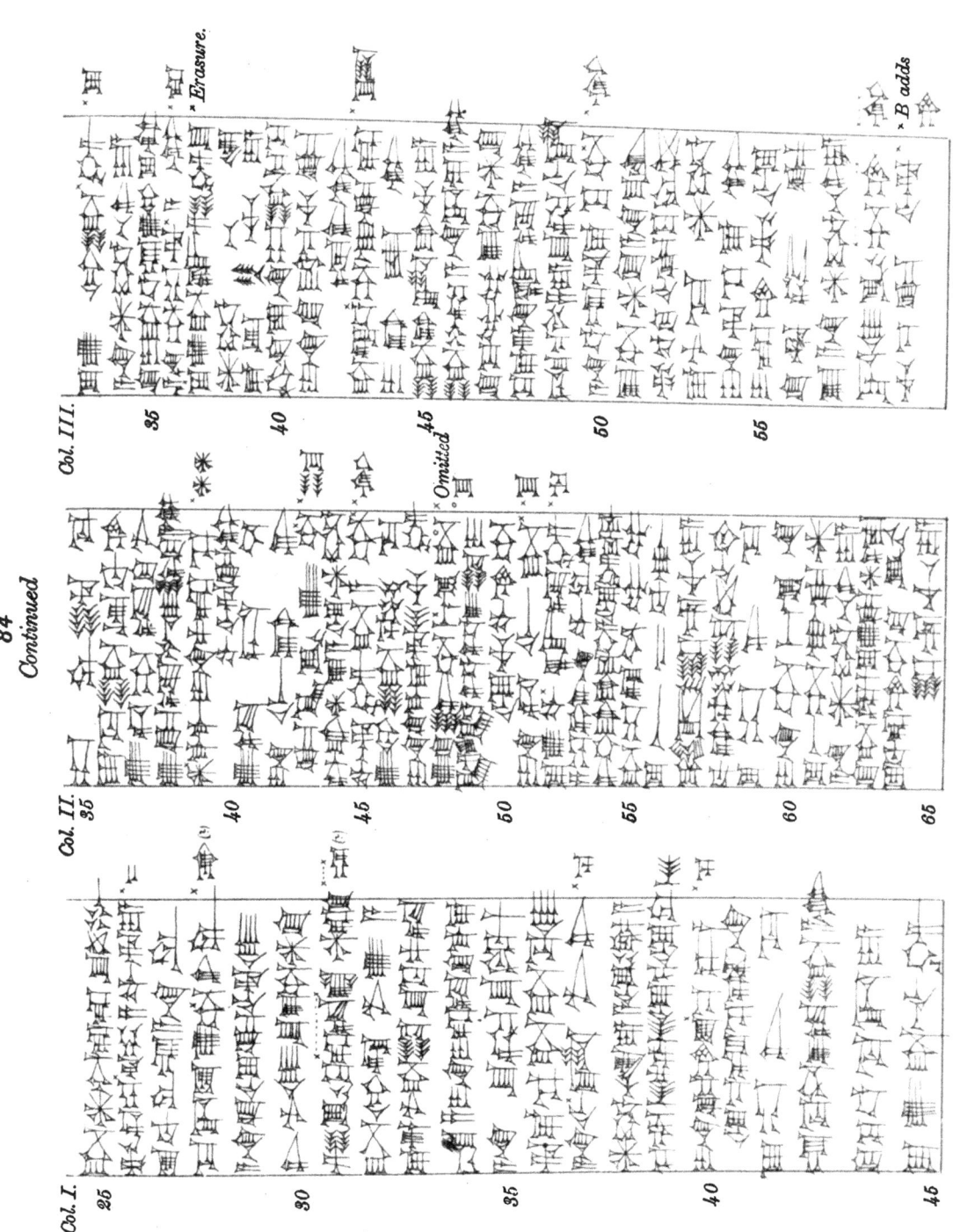

Pl. 34

85

Col. I.   Col. II.

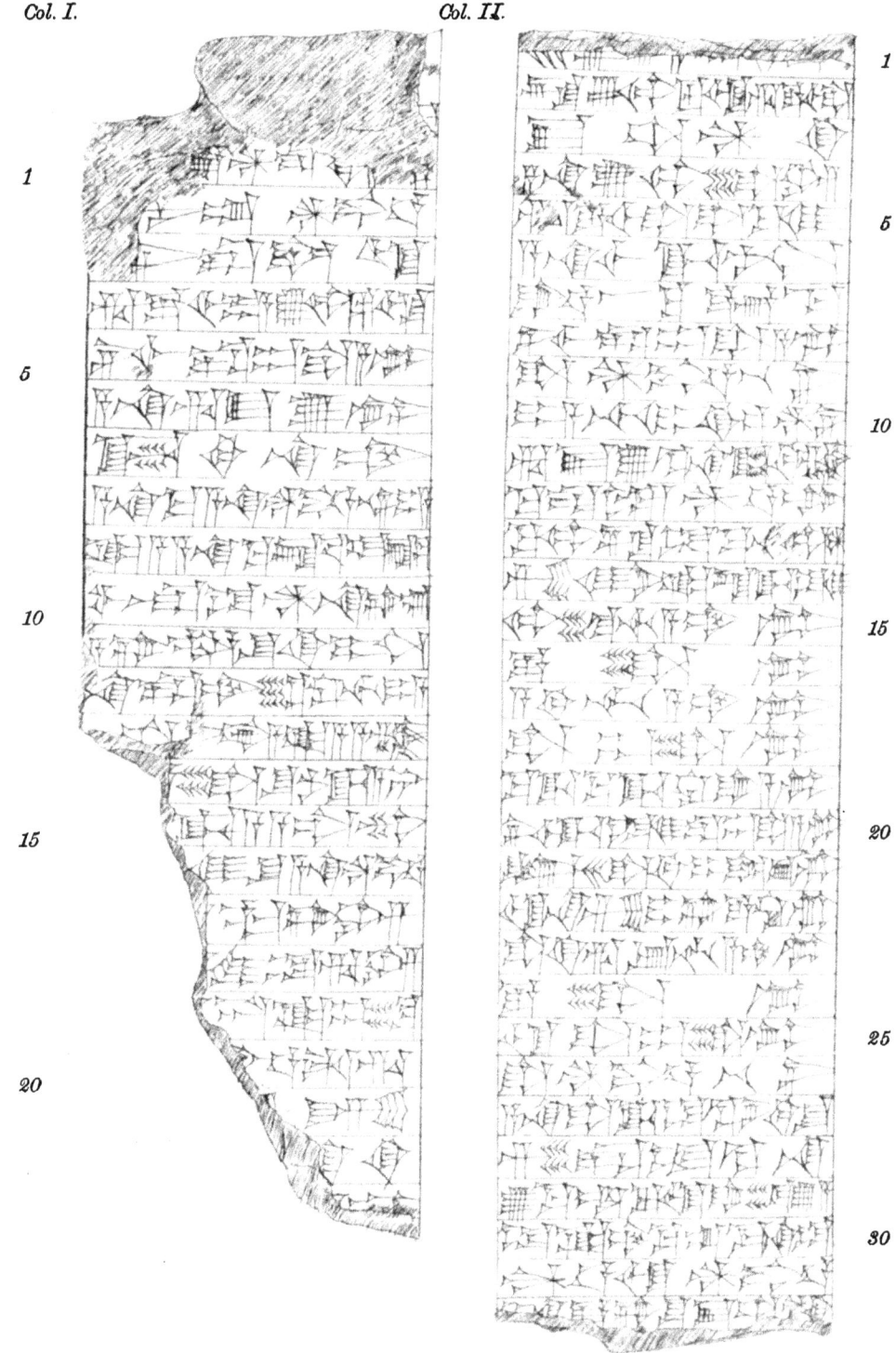

Pl. 35

85
Continued

Col. *III.*   Col. *IV.*

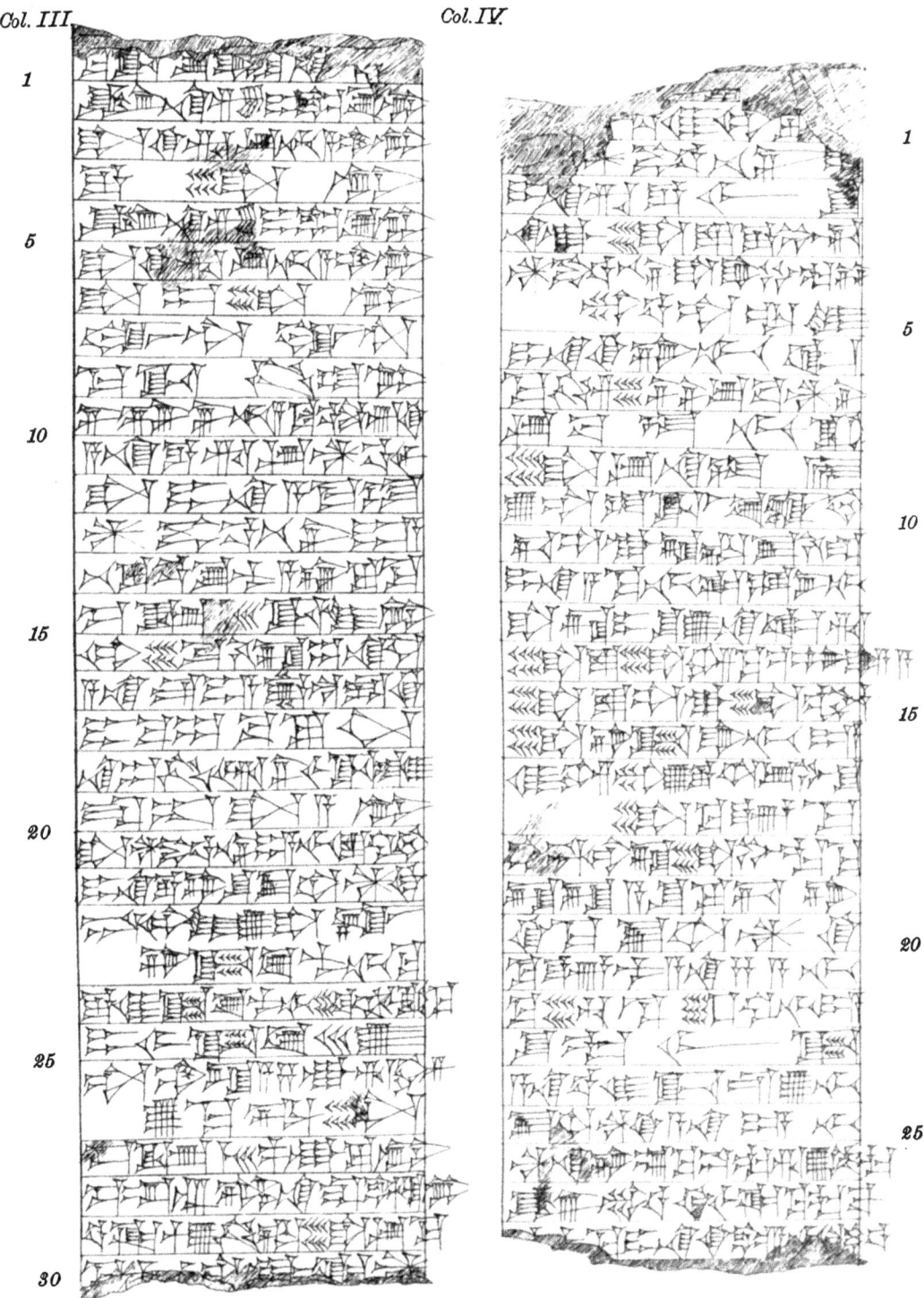

Trans. Am. Phil. Soc., N. S. XVIII, 1.

PL. I

1

DOOR-SOCKET OF SARGON I.
Nippur.

Trans. Am. Phil. Soc., N. S. XVIII, 1.

PL. II

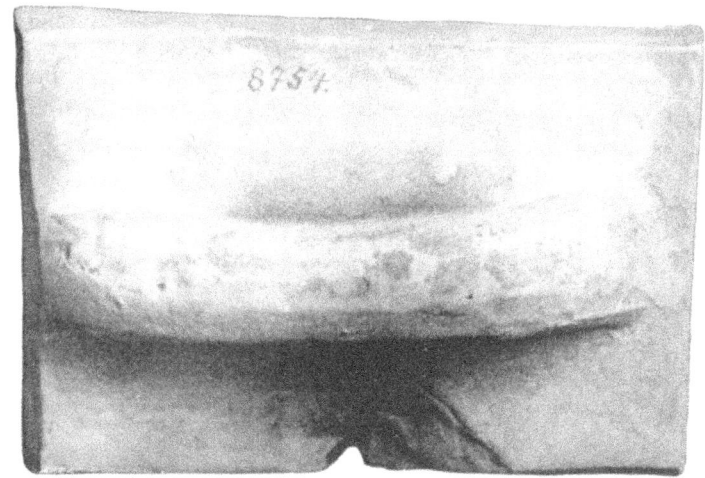

2

3

CLAY STAMPS FOR BRICKS.
Nippur.

2. Sargon I, Reverse.   3. Narâm Sin, Obverse.

Trans. Am. Phil. Soc., N. S. XVIII, 1.

PL. III

VASE FRAGMENTS OF ALUSHARSHID (URU-MU-USH).
Nippur.

Trans. Am. Phil. Soc., N. S. XVIII, 1.

PL. V

14

VASE FRAGMENT OF ALUSHARSHID (URU-MU-USH).
Nippur.

Trans. Am. Phil. Soc., N. S. XVIII, 1.

PL. VI

15

FRAGMENT OF A MARBLE SLAB: OBVERSE.
Abu Habba.

16

FRAGMENT OF A MARBLE SLAB: REVERSE.
Abu Habba.

18

17

19

17. FRAGMENT OF A MARBLE SLAB: EDGE - Abu Habba.

18, 19. Tablets of Baked Clay—Yokha.

Trans. Am. Phil. Soc., N. S. XVIII, 1.

PL. IX

20

21

20. STAMP OF HAMMURABI.   21. MORTAR OF BURNABURIASH.

Northern Babylonia.

Trans. Am. Phil. Soc., N. S. XVIII, 1.

PL. X

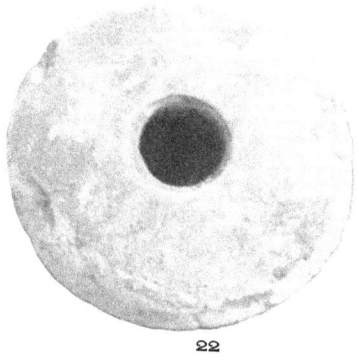

22

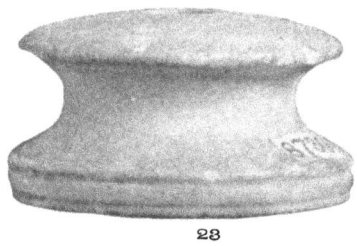

23

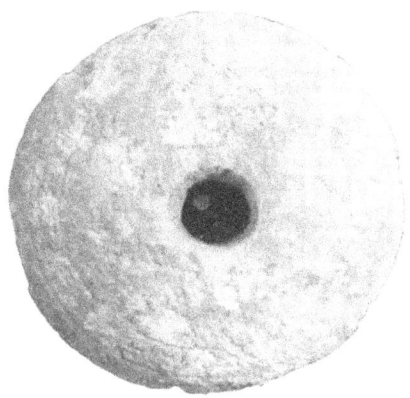

24

KNOBS OF SCEPTRES—Nippur.

22, 24. Magnesite (top view), Nazi-Maruttash.   23. Ivory (side view), Burnaburiash.

VOTIVE OBJECTS IN LAPISLAZULI AND IMITATION.
Nippur.

Trans. Am. Phil. Soc., N. S. XVIII, 1.

PL. XII

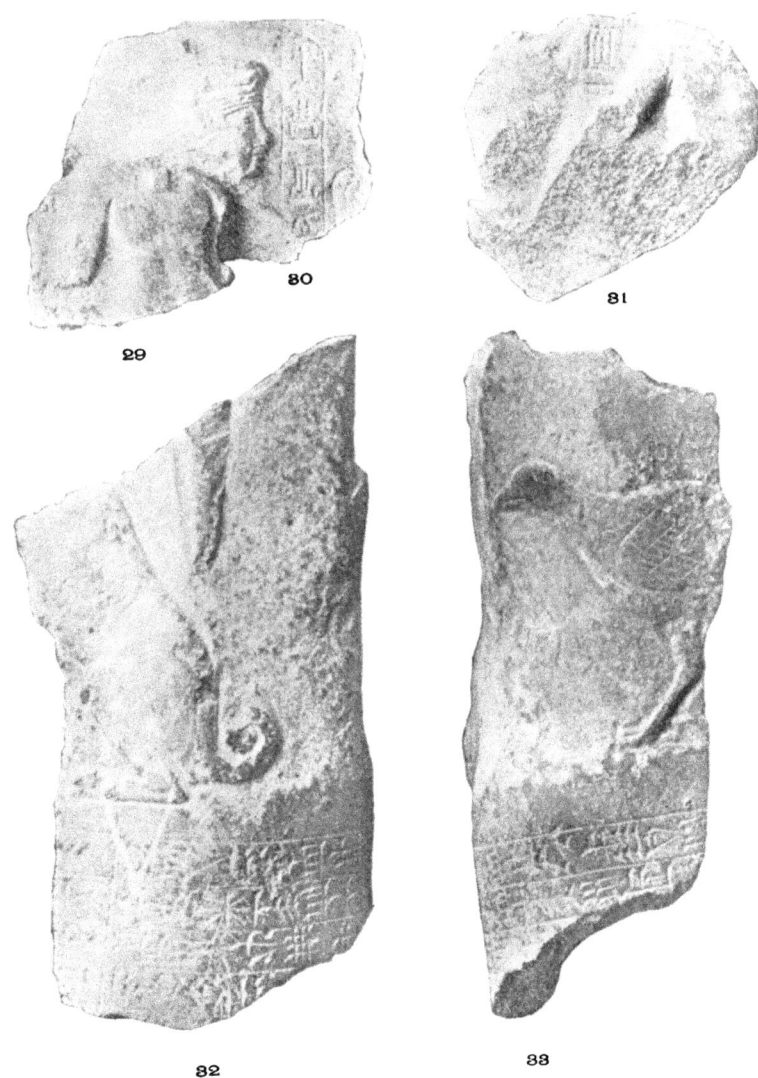

FRAGMENTS OF INSCRIBED BAS-RELIEFS.
Nippur.

Trans. Am. Phil. Soc., N. S. XVIII, 1.

PL. XIII

84

POINTED CLAY CYLINDER OF NABOPOLASSAR
Babylon.

85

BARREL-SHAPED CLAY CYLINDER OF NEBUCHADREZZAR II.
Babylon.

Trans. Am. Phil. Soc., N. S. XVIII, 1.

PL. XV

86

PLAN OF THE FIRST YEAR'S EXCAVATIONS AT NIPPUR.

The Roman numbers indicate the places where excavations were made ; the Arabic, the height of the mounds, in metres, above the present level of the canal bed. About five metres must be added to obtain the actual height above the plain. III Ekur—Bint el-Amir (Temple). VII Nimit-Marduk (Wall).

www.ingramcontent.com/pod-product-compliance
Lightning Source LLC
Chambersburg PA
CBHW080422190526
45161CB00004B/258

9781422377703